ANIMAL EXPERIMENTATION:

GOOD OR BAD?

Institute of Ideas
Expanding the Boundaries of Public Debate

DEBATING MATTERS

ANIMAL EXPERIMENTATION:

GOOD OR BAD?

Institute of Ideas
Expanding the Boundaries of Public Debate

Tony Gilland
Mark Matfield
Tom Regan
Stuart Derbyshire
Richard Ryder

Hodder & Stoughton
A MEMBER OF THE HODDER HEADLINE GROUP

DEBATING MATTERS

Orders: please contact Bookpoint Ltd, 130 Milton Park, Abingdon, Oxon OX14 4SB. Telephone: (44) 01235 827720. Fax: (44) 01235 400454.
Lines are open from 9.00 – 6.00, Monday to Saturday, with a 24-hour message answering service. Email address: orders@bookpoint.co.uk

British Library Cataloguing in Publication Data
A catalogue record for this title is available from the British Library

ISBN 0 340 85733 1

First published 2002
Impression number   10 9 8 7 6 5 4 3 2 1
Year   2007 2006 2005 2004 2003 2002

Typeset by Transet Limited, Coventry, England.
Printed in Great Britain for Hodder & Stoughton Educational, a division of Hodder Headline Plc, 338 Euston Road, London NW1 3BH by Cox & Wyman, Reading, Berks.

CONTENTS

DEBATING MATTERS

PREFACE

Since the summer of 2000, the Institute of Ideas (IoI) has organized a wide range of live debates, conferences and salons on issues of the day. The success of these events indicates a thirst for intelligent debate that goes beyond the headline or the sound-bite. The IoI was delighted to be approached by Hodder & Stoughton, with a proposal for a set of books modelled on this kind of debate. The *Debating Matters* series is the result and reflects the Institute's commitment to opening up discussions on issues which are often talked about in the public realm, but rarely interrogated outside of academia, government committee or specialist milieu. Each book comprises a set of essays, which address one of four themes: law, science, society and the arts and media.

Our aim is to avoid approaching questions in too black and white a way. Instead, in each book, essayists will give voice to the various sides of the debate on contentious contemporary issues, in a readable style. Sometimes approaches will overlap, but from different perspectives, and some contributors may not take a 'for or against' stance, but simply present the evidence dispassionately.

Debating Matters dwells on key issues that have emerged as concerns over the last few years, but which represent more than short-lived fads. For example, anxieties about the problem of 'designer babies', discussed in one book in this series, have risen over the past decade. But further scientific developments in reproductive technology, accompanied by a widespread cultural distrust of the implications of these developments, means the debate about 'designer babies' is set to continue. Similarly, preoccupations with the weather may hit the news at times of flooding or extreme weather conditions, but the underlying concern about global warming and the idea that man's intervention into nature is causing the world harm, addressed in another book in the *Debating Matters* series, is an enduring theme in contemporary culture.

At the heart of the series is the recognition that in today's culture, debate is too frequently sidelined. So-called political correctness has ruled out too many issues as inappropriate for debate. The oft noted 'dumbing down' of culture and education has taken its toll on intelligent and challenging public discussion. In the House of Commons, and in politics more generally, exchanges of views are downgraded in favour of consensus and arguments over matters of principle are a rarity. In our universities, current relativist orthodoxy celebrates all views as equal, as though there are no arguments to win. Whatever the cause, many in academia bemoan the loss of the vibrant contestation and robust refutation of ideas in seminars, lecture halls and research papers. Trends in the media have led to more 'reality TV', than TV debates about real issues and newspapers favour the personal column rather than the extended polemical essay. All these trends and more have had a chilling effect on debate.

But for society in general, and for individuals within it, the need for a robust intellectual approach to major issues of our day is essential. The *Debating Matters* series is one contribution to encouraging contest about ideas, so vital if we are to understand the world and play a part in shaping its future. You may not agree with all the essays in the *Debating Matters* series and you may not find all your questions answered or all your intellectual curiosity sated, but we hope you will find the essays stimulating, thought-provoking and a spur to carrying on the debate long after you have closed the book.

Claire Fox, Director, Institute of Ideas

NOTES ON THE CONTRIBUTORS

Stuart Derbyshire is an Assistant Professor in the Department of Anaesthesiology at the University of Pittsburgh in the United States. The University of London awarded Stuart his Ph.D. in 1995 following a four-year study of chronic pain patients using positron emission tomography (PET) at the Hammersmith Hospital in West London. He has continued his research into the central mechanisms of pain at the University of California, Los Angeles and at the University of Pittsburgh where he is currently a member of faculty. He has written several primary research reports as well as numerous essays exploring the psychological nature of pain and the validity of 'fetal pain'.

Tony Gilland is the Science and Society Director at the Institute of Ideas. He co-directed the Institute of Ideas' and New School University's Science, Knowledge and Humanity conference in New York 2001 and the Institute of Ideas' and Royal Institution's Interrogating the Precautionary Principle conference in London 2000. He is the editor of the Institute's *What is it to be Human? – What Science Can and Cannot Tell Us* Conversation in Print (2001). Gilland holds a degree in Philosophy, Politics and Economics from the University of Oxford.

Dr Mark Matfield worked as a medical research scientist in the UK and USA before taking a post with the Medical Research Council overseeing the funding of research in UK universities. In 1988 he was appointed as the Executive Director of the Research Defence Society

(RDS), the main organization representing scientists in the public debate about the use of animals in medical research. He has written and lectured widely on this issue for the last ten years.

Tom Regan is Emeritus Professor of Philosophy at North Carolina State University in the United States and president of The Culture & Animals Foundation. He is universally recognized as the philosophical leader of the animal rights movement. With more than a score of books to his credit, including his monumental *The Case for Animal Rights* (1983), Regan has challenged every aspect of humanity's exploitation of non-human animals, calling for abolition, not reform.

Dr Richard D. Ryder worked as a scientist in animal laboratories in Britain and the United States before becoming an animal protection campaigner, using only peaceful methods. His book *Victims of Science: The Use of Animals in Research* (1975) played a key role in the successful campaign for new legislation in the EU and Britain. Ryder's concept of *speciesism* is widely debated in philosophical circles and is now in most dictionaries. Ryder was for 20 years a hospital psychologist and, as a past Chairman of the RSPCA Council, he is known as that organization's 'chief modernizer'. His most recent books are *Animal Revolution: Changing Attitudes Towards Speciesism* (2000) and *Painism: A Modern Morality* (2001).

INTRODUCTION
Tony Gilland

In recent years the issue of animal experiments, always controversial, has been put under the spotlight – not least because of the activities of animal rights activists. Many people will now have heard of Huntingdon Life Sciences (HLS), a UK-based company that conducts legally required safety tests for drugs and other products on animals. HLS is the largest company in the UK providing animal testing facilities. In 1997 undercover footage showing unprofessional behaviour and inappropriate animal handling at HLS laboratories in the UK and US was screened on television stations in both countries, resulting in rebukes and fines for the company from regulatory authorities and an overhaul of its management. Subsequently HLS and its financial backers were subjected to an array of campaign activities, intimidatory threats and actual acts of violence, all of which kept the company in the newspaper headlines. HLS's work became so controversial that the company was unable to retain the services of banks and other financial institutions that neither wanted to be associated with the controversy nor to deal with threats and intimidation. In July 2001 the British Government was forced to take the unprecedented step of providing HLS with banking services via the Bank of England because the company could not acquire them anywhere else.

It is not only the recent successes of campaigners in obstructing the work of scientists and companies involved with animal testing that has given the issue of animal experimentation renewed attention in the public eye. Scientific advances, particularly in the fields of genetics and xenotransplantation (the grafting of animal cells, tissues and organs into humans for medical purposes), have also highlighted moral questions about the use of animals for scientific research in the quest to improve human health. For example, mice can now be genetically engineered to have missing genes or additional genes. In this way, and because of certain similarities between the genetic make-up of mice and humans, mice can be engineered to mimic human conditions such as cancer or cystic fibrosis, thereby helping scientists understand and develop treatments for human disease. Whilst fascinating scientifically, and potentially rewarding for humans, such developments have raised the controversial question of increased numbers of animal experiments being conducted – similarly with developments in the field of xenotransplantation.

More generally, discussions about modern farming practices, campaigns against live animal exports and parliamentary debates about fox hunting have kept the issue of animal welfare high on the political agenda. Within these debates, an increasing emphasis on the importance that many in society attach to animal welfare is clearly discernable. For example, modern methods of animal husbandry, where animals are often kept in confined spaces, have attracted much attention and criticism. Many supermarkets now stress the importance that they attach to the animal welfare practices of their suppliers. The Royal Society for the Prevention of Cruelty to Animals (RSPCA) has recently launched its 'Freedom Food' campaign – a labelling and monitoring system that requires farmed animals to be kept in conditions that ensure 'freedom from fear, distress, hunger, thirst, discomfort, pain injury, disease' and 'freedom to express normal behaviour'. Numerous celebrities and politicians have endorsed the campaign. Fox hunting remains a big political issue, with ongoing attempts being made by the House of Commons to ban or restrict hunts in the face of opposition within the House of Lords. The Scottish Parliament banned fox hunting early in 2002. In this context, the trend is towards placing greater emphasis on the welfare of animals in relation to their use in scientific research.

The contributors to this book, Mark Matfield, Tom Regan, Stuart Derbyshire and Richard Ryder, have considerable experience and varied opinions on the issues pertinent to the moral and political questions raised by animal experiments. Collectively, they address four key questions and areas of debate:

- What benefits are derived from animal experiments and do these justify inflicting harm on animals?
- What is the nature of the relationship between humans and animals?
- Do animals experience pain?
- Do animals possess rights?

NECESSARY EVIL, HUMAN GOOD OR SIMPLY IMMORAL?

Mark Matfield and Stuart Derbyshire highlight some of the high-profile medical advances that have been attributed to scientific research dependent on animal experiments – from blood transfusion to kidney dialysis, insulin treatment for diabetes, various organ transplants and the vaccine for polio. However, as Matfield points out, the UK – widely referred to as a nation of animal lovers – has a long tradition of being concerned about the use of animals in scientific research. According to Matfield, until the introduction of general anaesthesia in the later 1860s there was a general aversion to experimenting on animals among the middle classes in general and also among many scientists.

Following the introduction of general anaesthesia the number of officially recorded animal experiments conducted in the UK increased from 250 in 1881 to 95,000 in 1910.

Today the number of animal experiments conducted is vastly higher, at just over 2.7 million in the year 2000 (although this figure is approximately half what it was throughout the 1970s). This increase, since the turn of the twentieth century, reflects the amount of scientific research conducted in modern society and the number of products, particularly drugs but also many food additives, for example, which require animal testing before they can be legally sold for human use.

While society broadly seems to endorse the need to conduct experiments on animals, many regard them as something of a 'necessary evil'. Necessary to advance medical research and alleviate human suffering, but something that should be kept to a minimum. The success of anti-vivisectionists in getting leading cosmetics manufacturers such as Boots, Avon, Max Factor, Rimmel, Revlon, Yardley and Estee Lauder to cease the testing of their products on animals is indicative of public sentiments on this issue.

The UK's regulatory requirements, as laid down in the 1986 Animals (Scientific Procedures) Act, place a great emphasis on the welfare of the animals used in experiments and on minimizing the number of tests conducted. Legislation requires a cost benefit assessment (the costs of potential animal suffering versus the potential benefits of the research) to be conducted before any research project involving animals can go ahead. Furthermore, the research institute or company where testing is to take place, the individuals involved and the specific project must all be licensed by the Government before the work can proceed.

Legislation in the UK reflects the principle of the three Rs – replace, reduce, and refine – defined by Russell and Burch in 1959. The idea of the three Rs is, wherever possible, to replace animal experiments with non-animal ones, reduce the number of animals used in an experiment, and to refine the experiment so as to minimize suffering, distress or harm. The three Rs are strongly advocated by Matfield in his essay and supported by many of those who conduct experiments. The three Rs are strongly objected to by both Regan and Derbyshire, though for diametrically opposed reasons. Regan argues, whatever the benefits of animal experiments (and he and Ryder dispute that they are as great as is usually claimed), experimenting on animals for human benefit is simply morally wrong: 'the only adequate response to vivisection is empty cages, not larger cages'. Derbyshire, on the other hand, argues that the three Rs, by focusing on animal welfare rather than good scientific practice, endorses the fundamental misconception of the relationship between humans and animals that is held by those who object to animal experiments full stop. He opposes this perspective for 'reinforcing a lowlife opinion of animal researchers and encouraging the notion that animal experiments are problematic'.

 **HUMANS AND ANIMALS**

How we view ourselves as human beings has an important impact on how we view our relationship to animals at a philosophical level. This, in turn, impacts on our political and practical approaches to the question of animal experiments, and upon the general issue of animal welfare.

One important development that has influenced thinking on the relationship between humans and animals has been the resurgence of interest in the biological roots of human behaviour. Major advances in genetics, neuroscience, evolutionary biology, psychology and artificial intelligence have reinvigorated the debate about human nature. For example, some Darwinists believe that the 'selfish gene' theory can help explain human behaviour. More generally, evolutionary psychology, with its emphasis on understanding the limitations that our evolutionary and biological origins place upon our behaviour today, has been popularized through the writings of best-selling authors such as Richard Dawkins, Steven Pinker and Matt Ridley. The eminent American biologist and science writer E. O. Wilson even goes so far as to argue that the social sciences and other humanities can be understood as 'the last branches of biology' – and there is considerably more sympathy for his argument today than when he first made it in 1975.

This focus on biological explanations of human behaviour influences the debate about animal experimentation on a number of levels. The more that humans are seen as ultimately reducible to their biological make-up, the more the notion that humans are simply sophisticated animals appears convincing. The questioning of the distinction between humans and animals leads to both a questioning of the validity of the way in which humans treat animals as inferior, and lends weight to the argument that animals experience pain in the same way that humans do. Also, by raising questions about the extent to which humans actually make conscious choices, rather than merely thinking that they do, this approach has important implications for how we view human subjectivity. This, in turn, influences the discussion of what rights are and whether animals possess them.

The arguments put forward by Regan and Ryder demonstrate the importance of this debate to our understanding of the distinctions that can be legitimately drawn between humans and animals. Both refer to

Charles Darwin's work on evolution to argue that, with regard to the question of welfare, humans and animals are morally equivalent. According to Ryder: 'Thanks to Darwin, many of the huge and self-proclaimed differences between humans and animals were revealed to be no more than arrogant delusions. Surely, if we are all related through evolution we should also be related morally.' Similarly, Regan, in common with the influential philosopher Peter Singer, argues that when animals have a central nervous system it is reasonable to believe that they are capable of experiencing pleasure and pain in much the same way as we do and that therefore their enjoyment and suffering should be accorded equal status to our own.

By contrast, Matfield maintains a distinction between human suffering and that of animals. While he emphasizes that good scientific practice should, and does, coincide with a humane concern for the welfare of the animals subjected to the experiments, he argues that as long as experiments are conducted to 'high ethical standards' they are 'entirely justifiable'. The Royal Society, the body representing the UK's top scientists, adopts a very similar position. From this perspective, a concern for the welfare of animals used in experiments implies adopting a humane attitude towards those animals, rather than treating them as on a par with humans in terms of their suffering.

PAIN AND ANIMALS

How we understand pain and suffering is central to the question of animal experiments. According to Matfield, 80 per cent of experiments conducted on animals fall into the category of 'mild' and approximate to the level of distress caused by a simple injection. But if animals experience pain in the same way as humans do, then the argument put forward by animal rights campaigners – that animal experimentation can be likened to torture, particularly in the case of those experiments classified as 'substantial' – would carry a great deal of weight.

So do animals experience pain in a similar way to humans? According to Regan mammals share with humans 'a family of cognitive abilities' such as the ability to learn from past experience, remember the past and anticipate the future. They clearly 'enjoy some things and find others painful' otherwise they would not behave in one way rather than another to achieve pleasurable outcomes and avoid unpleasant ones. On this basis, Regan argues that although mammals 'lack the ability to read, write, or make moral choices', what they experience and what they are deprived of 'matters to them'.

Ryder, who coined the term 'speciesism' in the early 1970s, agrees with this viewpoint and argues that 'to discriminate against others merely because they have a different physical appearance is very unintelligent. Such speciesism is as irrational as sexism or racism.' Ryder supports the idea that animals experience pain by pointing to scientific advances in our understanding of biology. He argues that recent biochemical evidence 'indicates that in the nervous systems of all mammals, reptiles, birds and fish, are the same types of chemicals which are associated with the transmission and natural control of pain in our own'.

Scientific discussion about whether various physiological characteristics displayed by animals subjected to different stimuli can be interpreted as providing evidence of pain is a highly controversial area. Derbyshire argues that human pain cannot be understood simply in terms of physiology, and that we have a tendency to project human feelings onto animals. According to Derbyshire, the 'behaviour we observe following a noxious insult to an animal is disturbing to us because we do have insight and we project our expectation of feeling onto the animal.' He distinguishes humans from animals by the fact that we are conscious of being conscious and animals are not. He also argues that pain only means something in the context of the human social world and is not a purely physiological phenomena: 'the contents of our inner world come to mean something to us only in so far as they mean something to others ... if pain were an entirely private affair, no words would be able to express it because no external frame of reference would be comparable to and therefore adequate to express the sensation'. By contrast, he regards animals as 'mechanical' and 'driven by the dictates of nature'. According to this argument, the actions that animals take to avoid unpleasant experiences can be interpreted as mechanical and purely physiological processes. The same connotations cannot be attached to their experiences as ours. Therefore, animals cannot be said to experience 'pain'.

WELFARE AND RIGHTS

The question of whether animals possess rights, which provide them with a moral protection against being violated by humans, is a vexing one. Central to the question is the issue of whether one has to be capable of exercising rights in order to possess them.

One perspective on this question is to regard rights as freedoms that have to be fought for, won and defended, and to draw a distinction between the bearing of rights and the administration of welfare. From this perspective, the right to free speech, to freedom of association, to vote or to equal treatment under the law are rights that have at one time or another been fought for and won by different groups of people within society. In other words, the rights that exist within society do not have a pre-ordained basis in religion, moral philosophy or nature, but are the outcome of a political process.

Derbyshire adopts this perspective, drawing a distinction between welfare and rights. He argues that animals lack the ability to exercise rights, let alone argue for or fight for them, and therefore cannot possess them. Rather, according to Derbyshire, what is popularly discussed as 'animal rights' are choices that can be made by humans to protect animals – and the choices we make reflect our opinions and sentiments about our relationship to animals.

By contrast, Regan and Ryder do not draw this distinction between rights and welfare, but argue that the question of animal rights is a moral one. According to them animals have a basic moral right to bodily integrity and life, just as humans do. From this perspective it is logically inconsistent to uphold the rights of humans to bodily integrity and life and not those of animals. Ryder's philosophical perspective also places a great deal of emphasis on the avoidance of pain and suffering as the most important moral issue of our times. He argues that 'conditions such as justice, equality and liberty are all morally subordinate to the reduction of pain' and that they are important in relation to the extent that they help to alleviate pain and suffering.

From their very different perspectives, the essays in this book provide important insights into and arguments about the question of animal experimentation. They also provide fascinating reflections upon the fundamental question of what it is to be human. We hope you enjoy them.

Essay One

ANIMAL EXPERIMENTS: MEDICAL PROGRESS AND ANIMAL RIGHTS PROTESTS

Mark Matfield

Animal experiments have played a vital role in advancing our knowledge of human and animal biology. They have been crucial to a host of medical advances and the development of treatments for human and animal disease. However, the issue of animal welfare has always been of concern to both scientists and the public. In the UK we have a strong tradition of caring for the welfare of animals used in laboratory experiments and stringent regulations that govern every aspect of animal research and testing. Whilst the public are concerned about the welfare of laboratory animals, and expect experiments to be conducted humanely and only when necessary, they are also supportive of the need for animal experiments to take place. In this essay, I will outline the history of animal experiments and the achievements associated with these, alongside the history of care for animals in laboratories. I will argue that animal experiments are vital to the future well-being of humans and, as long as they are conducted to high ethical standards, they are entirely justifiable.

THE HISTORY OF ANIMAL EXPERIMENTATION

Animal experiments have been used in medical science since its origins in ancient Greece. Around 500 BC Alcmaeon of Croton demonstrated the function of the optic nerve by cutting through it in living animals and showing that blindness resulted. Herophilus of

Alexandria studied animals between 330 BC and 250 BC and showed the different functions of nerves and tendons (N. A. Rupke, (ed.), *Vivisection in Historical Perspective*, Croon-Helm, 1987).

The scientific and ethical reasons that animals were used for these rudimentary experiments of ancient times are essentially the same today. Firstly, it was apparent to the doctors of those times that they did not know enough about how the body worked to understand why it malfunctioned when diseased and that there was sufficient biological similarity between humans and other animals that a finding in one could be applied with confidence to the other. The Greeks also valued the pursuit of knowledge for its own sake. Secondly, even though human suffering was commonplace and human life held cheap in those days, it was still considered far more important than the life of a non-human animal. However, there was one major difference. The degree of cruelty and suffering involved would have been intolerable by modern standards. These experiments were performed on fully conscious animals.

The Greek tradition of studying animals to help understand the human body continued into the Roman era and into the Arabic schools of medicine, but died away with the coming of the Dark Ages. It was not until the sixteenth century that it was revived in the medical schools of Italy, whence it spread throughout Europe as the usefulness of experiments on living animals to study the function of the body organs became clear. Many of the most fundamental discoveries in physiology came from studying animals, including William Harvey's demonstration of blood circulation in 1628, Robert Hooke's discovery of the function of the lungs in 1667 and Stephen Hales's measurement of blood pressure in 1733 (P. Rhodes, *An Outline of the History of Medicine*, Butterworths, 1985).

However, as the study of animals advanced in medical schools across Europe and the experiments became more complex and invasive, doctors and scientists in the UK began to display a growing reluctance to cause animal suffering in the name of science. They appeared to share the growing sentiment towards animal welfare that was found amongst the predominantly urban British middle classes. In 1863, an editorial about animal experiments in the leading medical journal, *The Lancet*, stated, '… perhaps some two or three, or at most six, scientific men in London are known to be pursuing certain lines of investigation which require them occasionally during the course of a year to employ living animals for the purpose of their inquiries.' As a result, experimental physiology – the main type of medical research in those times – was quite underdeveloped in the UK by comparison with the rest of Europe.

It was not until the introduction of general anaesthesia in the late 1860s that things began to change in the UK and a new generation of young medical scientists began to conduct research on animals rendered unconscious with ether or chloroform. Government statistics show that the number of animal experiments conducted in the UK increased from 250 in 1881 (the first year that records were kept) to 95,000 in 1910.

The promise of animal research began to pay off during the twentieth century when much greater investment in the biological and medical sciences produced a remarkable number of medical advances, essentially all of which depended, at one or more point in their development, on animal experiments. Before 1922, juvenile diabetes meant a slow, painful death as the body was starved of nutrition with no cure and no sufferer surviving beyond the age of 20. The discovery of insulin, through research on dogs and rabbits, came as a virtual miracle. For the first time, a major fatal disease could be cured, simply by regular injections. It is estimated that, throughout the world, up to 30 million lives have been saved by this discovery. It came as no surprise to the world when, in 1923, Frederick Banting and John MacLeod were awarded the Nobel Prize for Medicine for their work in this field.

The story of the discovery of penicillin by Alexander Fleming is well known to many. He noticed that some fungus growing on an agar plate somehow prevented bacteria growing near it. In 1929 he isolated the substance responsible from the Penicillium fungus and showed that it had the ability to stop bacterial growth. However, it is less well known that Fleming then put this line of research aside, before conducting the crucial animal test for antibiotic activity. Many things, including bleach and sulphuric acid, will kill bacteria. The essential thing about antibiotics is that they kill bacterial cells but do not harm animal cells. Fleming failed to perform the crucial animal experiment that would have shown this. It was not until 1940, when Ernst Chain and Howard Florey showed that injections of penicillin protected mice that had been given an otherwise lethal dose of virulent bacteria, that the full potential of penicillin was discovered. The discovery was recognized in 1945 with the award of the Nobel Prize jointly to Fleming, Chain and Florey.

The development of vaccines against polio depended heavily upon animal experiments. Although the disease was known from ancient times, it was not until 1909 that we began to understand how it was caused, when researchers studying the disease in monkeys (the only group of animals affected by it) showed that it was transmitted by a tiny particle, smaller than bacteria. From then until 1949, when John Enders and Thomas Weller were able to grow the virus in tissue culture, the only way to detect it was by infecting animals: originally monkeys and later mice. In the early 1950s Professor Albert Sabin created vaccines against polio by growing the virus in mice, rats, monkeys and tissue culture. In 1947, the number of polio cases in the UK hit a high with nearly 8,000 people becoming paralysed for life or dying as a result of contracting the disease. The vaccination programme started in 1958 and the number of cases dropped to below 100 within six years. There has been only one case of polio in the UK since 1990 and that was contracted while the person was travelling abroad.

The list of medical advances that have depended on animal research (see Table 1) is a matter of medical history. It includes essentially every important medical advance of the twentieth century. The total saving of human life and amount of suffering prevented is impossible to calculate. Without the developments that depended on animal research, our medical system would differ little from that of the late-Victorian period. At the time of these medical advances, the use of animals was essential for the research that produced them. Even with present-day scientific technology, which has greatly reduced the numbers of animals needed for research, it is difficult to see how most of these experiments could be carried out without using animals.

THE RISE OF LABORATORY ANIMAL WELFARE

The very first proposals to safeguard the welfare of animals used in experiments were made by a British animal experimenter. In 1831, Marshall Hall elaborated five principles which included: not using animals where the results could be obtained by another method, proper experimental design, avoiding the unnecessary repetition of animal experiments, minimizing any suffering inflicted upon the animal, and the proper recording of results so as to remove any necessity for repetition (N. A. Rupke, (ed.), *Vivisection in Historical Perspective*, Croon-Helm, 1987). It seems unlikely that Hall developed these principles out of altruism alone. He was carrying out experiments on animals in London in pre-anaesthesia times and received a significant level of criticism both from within the medical and scientific professions and from other members of London society.

The United Kingdom was the first country in the world to have laws regulating the use of animals in experiments. Unfortunately, the debate that led up to that legislation was heated and very divisive. The main provisions of the original Bill, which were extremely restrictive, were based on proposals by an anti-vivisection organization. The medical and scientific community lobbied hard and secured a number of changes before it was passed as the Prevention of Cruelty Act of 1876 (R. D. French, *Antivivisection and Medical Science in Victorian Society*, Princetown University Press, 1975).

One of the unfortunate effects of this controversy was that the scientists resented being regulated and saw animal welfare as something that got in the way of animal research. However, with time, these attitudes have changed. In 1986 new and more effective legislation about animal experimentation was introduced with the Animals (Scientific Procedures) Act 1986. The prevailing view among animal researchers these days, is that good animal science and good animal welfare go hand in hand. Not only is it considered ethically wrong to have poor laboratory animal welfare; it is also regarded as bad science.

The forefathers of laboratory animal welfare are two UK scientists, Bill Russell and Rex Burch. In 1958 the Universities Federation for Animal Welfare awarded them fellowships to study the ethical aspects of animal experimentation. Their seminal book *The Principles of Humane Experimental Technique* (Methuen, 1959), defined the principle of the three Rs – Replace, Reduce and Refine – as the basis for humane experimental procedures in animal research. The three Rs

Table 1 Medical progress that depended on animal experiments

Date	Medical advance	Animals experimented on
1910s	Blood transfusion	Dogs
	Kidney dialysis	Rabbits, dogs
1920s	Insulin to treat diabetes	Rabbits, dogs
1930s	Modern anaesthetics for surgery	Rats, rabbits, dogs, monkeys
	Diphtheria vaccine	Guinea pigs
1940s	Broad-spectrum antibiotics for infections	Mice, rats, hamsters, rabbits, dogs
	Whooping Cough vaccine	Mice, rabbits
	Heart-lung machine for open-heart surgery	Dogs, pigs
1950s	Kidney transplants	Mice, cats, dogs
	Cardiac pacemaker	Dogs
	Polio vaccine	Mice, monkeys, rabbits
	Medicines for high blood pressure	Mice, rats, rabbits, cats, dogs
	Joint replacement surgery	Rabbits
1960s	Medicines for mental illnesses	Rats, rabbits
	Replacement heart valves	Guinea pigs, rats
	Coronary bypass operations	Pigs, dogs
	Heart transplants	Pigs, dogs
1970s	Drugs to treat ulcers	Rats, dogs
	Improved sutures and surgical materials	Rabbits, rats
	Drugs to treat asthma	Guinea pigs, rats, rabbits
1980s	Immunosuppressant drugs for organ transplants	Mice, rabbits, dogs
	CAT scanning for improved diagnosis	Pigs
	Life-support systems for premature babies	Monkeys
	Medicines for virus diseases	Mice
1990s	Meningitis vaccine	Mice, rabbits
	Gene therapy for cystic fibrosis	Mice
	Electronic implants for deafness	Ferrets

have now become the ethical basis for laboratory animal science and welfare throughout the world.

Replacing animal experiments with non-animal techniques wherever possible is important, not only ethically but often also for scientific reasons. Many non-animal techniques are faster, cheaper and more able to give useful data than animal techniques. In fact, the majority of medical research, perhaps 80–90 per cent, is carried out using non-animal techniques. However, these are not alternatives to animal experiments. To discover how cancer cells are different from normal cells, we have to study the cells in isolation. The same information cannot be discovered using an animal experiment. Normally, non-animal techniques are used in experiments because they can reveal information that can only be found using those particular techniques. Thus, non-animal techniques tend to complement rather than replace animal techniques of experimentation. New techniques are developed to discover new information. Sometimes a new non-animal technique produces enough information of the right type that it is no longer necessary to do the animal experiment. In this way, new non-animal methods that were developed simply to be useful in themselves can also replace animal experiments. Deliberately developing techniques as replacements for existing animal techniques has proven more difficult.

Reducing the numbers of animals used in each experiment to the minimum necessary to get meaningful results is the second of the three Rs. It is sometimes, incorrectly, described as reducing the overall number of animal experiments. This is determined by other factors, such as the amount of research being done. However, within a single experiment, statistical calculations can define the minimum number of animals that need to be used to give reliable results. To keep this number to a minimum, it is necessary to use the proper experimental design and to eliminate any other factors that might influence the results. A lot of research uses SPF (Specific Pathogen Free) animals, which are free from common infections and have been bred in sterile conditions. By eliminating the risk of past or present infections, which might influence the outcome of the experiment, it is possible to use fewer animals to get reliable results.

Refining an experiment (the third R) means to design and carry it out in such a way as to reduce any suffering, distress or harm to the animal to a minimum. In many ways this is the most important of the three Rs. The simplest application of refinement is to give the animal an anaesthetic or pain-killer if that will reduce its suffering. A more thorough application of the principle includes consideration of the whole life of the animal. Since almost all laboratory animals are specially bred, they spend the majority of their lives not being experimented upon, but simply living in the laboratory animal house. They could live alone in a barren cage or they could live in social groups, in large enclosures that mimic the complex environment they would experience in the wild. Research into animal welfare has conclusively proven that animals deprived of stimulation (particularly social stimulation) experience significant stress and develop behavioural abnormalities. The last 20 years has seen a steady trend towards environmental enrichment for laboratory animals. Sometimes this can be remarkably simple. For years, monkeys were kept in individual cages, 20 or so to a room. Some places now keep them in the same rooms, with the same cages, but with the doors open. Instead of each animal being isolated in a cage, the animals now live in a social group, in a room full of structures to explore, hide in or climb over. The addition of a few ropes and branches enriches their environment even further.

An important part of refinement is to establish and use the most humane end-point. For example, an experimental technique used to determine if a substance causes cancer might have originally involved administering the substance and then counting how may animals die of cancer over the next few months. However, it is rarely necessary to use the death of an animal as the end-point in the experiment. Instead, all the animals could be – and normally are – put down painlessly before any have got advanced cancers, and the number of cancers per animal measured at autopsy. The aim is not to let any of the cancers get much beyond the size at which they can be first detected. This still gives a reliable measure of the substance's potential to cause cancer, but it does not involve any animals dying from or even suffering as a result of the cancers.

The UK system of regulating animal experimentation places a considerable emphasis on the three Rs. Every animal experiment is controlled by law and counted in statistics that are published by the Government. Every experimental procedure to be used on any animal has to be explained and justified to government Inspectors. Each procedure is categorized according to the maximum degree of suffering it might cause. *Mild* means the procedure can cause no more distress than would a simple injection. *Substantial* procedures include surgical operations and similar techniques that have the potential to cause serious suffering. *Moderate* procedures are the broad category between Mild and Substantial.

For every research project, the numbers of procedures of different severity are assessed and the project is given an overall rating of Mild, Moderate or Substantial. A small percentage are unclassified: these are when all the animals are given a general anaesthetic for the whole time and put down whilst still unconscious. The numbers of projects in each category are counted and published. In 2000, 40 per cent of projects were mild, 55 per cent moderate and two per cent substantial (*Statistics of Scientific Procedures on Living Animals Great Britain 2000*, The Stationary Office,

CM 5244). However, a project can be classified as Substantial even when only a quarter or a third of the individual procedures fall in that category. The remainder would be Moderate or Mild.

The number of procedures (that is, individual experiments) in each category is not recorded. However, experts estimate that over 80 per cent of all procedures in the UK are now Mild – they cause no more distress than a simple injection.

PUBLIC ATTITUDES TOWARDS ANIMAL EXPERIMENTS

It is easy to assume that the general public are opposed to animal experimentation and they think no further about it. However, this assumption is wrong. In recent years there have been some detailed and careful opinion surveys on this subject in the UK, which revealed a more complex picture of public attitudes (A. Coghlan *et al.*, 'Let the people speak', *New Scientist* 22 May 1999 and B. Davies, 'In-depth survey of public attitudes shows surprising degree of acceptance', *RDS News* April 2000, 8–11).

Both surveys confirmed that a substantial majority of the public accepted, in principle, that it was necessary to use animals in medical research. However, their acceptance was dependent upon a number of conditions. These conditions related to the purpose of the research, which they wanted to be into serious medical conditions; the type of animal used, where they were more willing to accept the use of mice than monkeys; and the degree of pain or suffering involved, which they wanted to see minimized or eliminated entirely. In fact, those three conditions are some of the main provisions of the system of regulating animal research in the UK.

One of the surveys extended this analysis by asking people what they would want to see in a system that regulated animal experiments. In their unprompted responses they suggested all of the key elements of the existing UK system of regulation. This suggests that if the public were more aware of the existing regulatory system, they might well have more support for the way animals are used in research in the UK. However, much of the public debate about animal experimentation has focussed on the question of whether or not it is necessary to use animals in research. This tends to obscure the issues about animal welfare and how the system is regulated, which seem to be discussed in the media much less often.

Public attitudes to animal experimentation have undoubtedly been shaped by a lack of trust of scientists in general and a lack of openness by the scientific community on this subject in particular. Public distrust of scientists has grown in recent years in the UK and many commentators have pointed to examples such as BSE, genetically modified crops and global warming as causes of this lack of trust. It is worth pointing out that placing the blame for these incidents on the scientific community is less than completely fair in that it ignores the rather greater role played by the Government and commercial interests.

It is quite true that there has been a general lack of openness about animal experimentation in the UK over the last 30 years or so. However, there has been a good reason for this. Animal researchers are very worried about being targeted by animal rights extremists. The Animal Liberation Front has deliberately attacked or threatened the few scientists who take a high public profile in support of animal research. It is reasonable to assume that their motive was to frighten other scientists away from speaking out in public about the use of animals in medical research. If this was their strategy, it has been largely successful.

ANIMAL RIGHTS: FROM CAMPAIGNING TO EXTREMISM

The antivivisection movement began in Britain nearly 150 years ago. Its objective was always clear – to end all animal experimentation – and its methods were legitimate. Typically the antivivisectionists would seek to influence public and political opinion through lectures, debates, leaflets, posters and other publications.

During the 1970s a new movement arose, which rapidly overtook all the antivivisection organizations. Animal rights is both a philosophy and a style of activism. The philosophy, first defined by Peter Singer, argued that non-human animals have equal or very similar rights to humans and thus we have no right to use other animals in any way: not for food, clothing, in research or even as pets. Animal rights activism began with the hunt saboteurs who used what they called *direct action* to disrupt fox and otter hunts. However, the movement that started with groups of determined activists putting themselves between the hunters and the hunted soon developed into something much closer to urban terrorism.

In 1972, several members of the Luton branch of the Hunt Saboteurs Association decided to take direct action to a different level. Their activities began with damage to the vehicles used by the local hunt and moved rapidly to arson attacks on the new research laboratories of a pharmaceutical company. It must be admitted that these early arson attacks were not sophisticated and it was not long before Ronnie Lee and Cliff Goodman were arrested when returning to the scene of one of their crimes to attempt another arson attack. They served a short prison sentence and, when released, Lee reformed his group of activists into the Animal Liberation Front (ALF).

In their early years, the ALF devoted a significant part of their activities to 'liberating' animals – stealing them from laboratories, breeders and farms. However, most of their actions consisted of criminal damage: trashing laboratories, vandalizing cars and arson attacks on office buildings. In time, the liberation activities were abandoned, the attacks became more violent and animal research became the favourite target. In 1982, ALF tactics took a new turn when letter bombs were sent to the leaders of the four main political parties, injuring a civil servant at Downing Street (D. Henshaw, *Animal Warfare*, Fontana/Collins, 1989).

This trend towards greater violence continued with a series of petrol bomb attacks on the homes of medical researchers in 1985. Later the same year, the first ALF car bombs were used. They were placed under the parked cars of two scientists working at a research institute in Surrey. The devices, which used a crude home-made explosive designed to be set off by timers, were placed on the ground under the cars, so one assumes that the intention was to destroy the vehicles rather than harm the drivers. However, an anonymous spokesman for the 'Animal Rights Militia', who claimed responsibility for the attacks, told the local paper 'We will go to any lengths to prevent these animal abusers' murderous activities. If that means killing an individual we will not shy away from such action.'

The next ten years were the heyday of animal rights extremism. Claimed in a variety of names (Animal Rights Militia, Animal Liberation Front, Animal Defence League, Animal Liberation League, and so on), their tactics included product contamination (of Mars Bars, Lucozade, frozen chickens and various toiletries), mass invasions and damage to various research facilities, arson attacks on shops (including one that burnt down an entire department store) and bombs set off in fast-food stores during the lunch hour.

The most extreme attacks took place in Bristol and Salisbury. One morning in February 1989, the *Daily Mirror* received a call from a spokesman from the Animal Abused Society claiming that a bomb would go off at noon in the Senate House building of Bristol University. The building was cleared and searched but nothing was found. No bomb went off at noon. However, at midnight a five-pound plastic explosive device destroyed an entire floor of the building. The police investigating the incident were confident that the device had been planted after the original police search, when people were let back into the building.

The same explosive, PE4, was used in two car bomb attacks on scientists in June 1990. Sophisticated devices with mercury-tilt switches and about half a pound of high explosive were attached to the underside of the cars of Dr Margaret Baskerville and Dr Max Headley. Both devices went off while the scientists were driving. The explosion buckled the frame of Dr Baskerville's jeep, jamming the doors and blowing out both rear and front windows. Fortunately, she managed to climb out of the front window before the car caught fire and escaped with only minor injuries. The device under Dr Headley's car did not go off for some time. If the devices were placed at around the same time, then Dr Headley must have driven the car around for four days with it attached. It went off when he was driving through Bristol, blowing a hole in the floor of his car. Miraculously, he too escaped any serious injuries. However, it seems clear that both devices were intended to kill. As it was, a baby boy over 20 feet away from the blast was seriously injured by the shrapnel from the Bristol explosion.

Although there was a large animal rights letter-bombing campaign during the mid-1990s, the Bristol and Salisbury car bombs appear to have been the peak of animal rights extremism in the UK. In the mid-

1990s a new style of animal rights extremism arose in campaigns against companies such as Huntingdon Life Sciences. This Cambridgeshire company carries out safety tests on medicines, food additives and other chemicals on behalf of other companies. The majority of these tests are required by law and many of them involve animals. In 1999, a group of radical animal rights activists launched a campaign to force the company out of business. From the outset the campaign was highly aggressive with staff being harassed and intimidated at home as well as at their work. Soon the harassment spread to include the company's shareholders, banks, stockbrokers and clients. It escalated to violence with fire-bombings of some employees' cars, parked outside their homes. In 2001, when the Managing Director returned home one evening, he was attacked and injured by three masked assailants wielding baseball bats.

Animal rights extremism became such a serious problem that, in early 2001, the Government rapidly introduced a series of laws and set up a special police squad to deal with this problem. Eventually, the ringleaders of the campaign were convicted and imprisoned for conspiracy to incite criminal damage. The extreme nature of the tactics used by some of the activists has stimulated public and political opinion in favour of animal research.

CONCLUSION – THE FUTURE OF THE DEBATE ABOUT ANIMAL EXPERIMENTS

Making predictions – particularly in print – is always a risky business. However, there do appear to be some clear trends in the public and political debate about animal experimentation and it would seem sensible to look at them and ask whether they will continue and where will they lead.

The first matter to consider is the current campaign against Huntingdon Life Sciences. At the time of writing – early 2002 – this campaign definitely appears to be running out of steam, and may well fade to nothing before the end of the year. In fact, there is a noticeable cycle in the tactics of animal rights extremism in the UK. Over a five to seven year period, particular tactics – be they arson attacks, letter bombs or focussed campaigns of harassment – arise, run their course and fade away. Then, after a gap of a year or so, a new tactic emerges. The tactics used in the campaign against Huntingdon Life Sciences emerged in late 1996. If things follow their usual course, there will be little animal rights extremist activity for a while and then there will be a new campaign, against a different target or targets and using different tactics.

One noticeable trend over the last decade is that there now appears to be a greater public acceptance that it is necessary to use animals in medical research and in developing better treatments. I expect that this situation will continue and, in general, there will be less public support for the antivivisection cause. It is noticeable that very few antivivisection organizations now run effective campaigns trying to persuade their audience that animal experimentation is worthless or unnecessary. Instead, they focus on secondary issues, such as lack of openness, or the justification for using a particularly emotive species such as primates or dogs. I suspect that this trend will continue, with overt antivivisection campaigning becoming less common.

Finally, I hope that there will be greater openness by the scientific community about animal research. Fear of being targeted by animal rights extremists has made the majority of animal researchers unwilling to engage in the public debate on this subject and reluctant to be more open. However, the most effective method of informing people about the animal welfare standards in our laboratories is simply to let them in to see for themselves. Seeing the animals themselves, the people who care for them and the ever-present concern for animal welfare running through all our animal experimentation is extraordinarily impressive, particularly if the visitor fears that they are going to see unpleasant scenes of animal suffering. Despite the activity of animal rights extremists, in recent years more and more medical researchers have been willing to speak out in public about the use of animals in their research. I hope this will soon be followed by even greater openness so that people can see how the welfare of animals is safeguarded in our laboratories.

Essay Two

EMPTY CAGES: ANIMAL RIGHTS AND VIVISECTION
Tom Regan

Animals are used in laboratories for three main purposes: education, product safety testing and experimentation (medical research in particular). Unless otherwise indicated, my discussion is limited to their use in harmful, non-therapeutic medical research (which, for simplicity, I sometimes refer to as vivisection). Experimentation of this kind differs from therapeutic experimentation, where the intention is to benefit the subjects on whom the experiments are conducted. In harmful, non-therapeutic experimentation, by contrast, subjects are harmed in the absence of any intended benefit for them; instead, the intention is to obtain information that might ultimately lead to benefits for others.

Human beings, not only non-human animals, have been used in harmful, non-therapeutic experimentation. In fact, the history of medical research contains numerous examples of human vivisection, and it is doubtful whether the ethics of animal vivisection can be fully appreciated apart from the ethics of human vivisection. Unless otherwise indicated, however, the current discussion of vivisection, and my use of the term, are limited to harmful, non-therapeutic experimentation using non-human animals.

THE BENEFITS ARGUMENT

There is only one serious moral defence of vivisection. That defence proceeds as follows. Human beings are better off because of vivisection. Indeed, we are *much* better off because of it. If not all, at least the majority of the most important improvements in human health and longevity are indebted to vivisection. Included among the advances often cited are open-heart surgery, vaccines (for polio and small pox, for example), cataract and hip replacement surgery, and advances in rehabilitation techniques for victims of spinal cord injuries and strokes. Without these and the many other advances attributable to vivisection, proponents of the Benefits Argument maintain, the incidence of human disease, permanent disability, and premature death would be far, far greater than it is today.

Defenders of the Benefits Argument are not indifferent to how animals are treated. They agree that animals used in vivisection sometimes suffer, both during the research itself and because of the restrictive conditions of their life in the laboratory. That the research can harm animals, no reasonable person will deny. Experimental procedures include drowning, suffocating, starving, and burning; blinding animals and destroying their ability to hear; damaging their brains, severing their limbs, crushing their organs; inducing heart attacks, ulcers, paralysis, seizures; forcing them to inhale tobacco smoke, drink alcohol, and ingest various drugs, such as heroin and cocaine.

These harms are regrettable, vivisection's defenders acknowledge, and everything that can be done should be done to minimize animal suffering. For example, to prevent overcrowding, animals should be housed in larger cages. But (so the argument goes) there is no other way to secure the important human health benefits that vivisection yields so abundantly, benefits that greatly exceed any harms that animals endure.

⬤⬤◆ WHAT THE BENEFITS ARGUMENT OMITS

Any argument that rests on comparing benefits and harms must not only state the benefits accurately; it must also do the same for the relevant harms. Advocates of the Benefits Argument fail on both counts. Independent of their lamentable tendency to minimize the harms done to animals and their fixed resolve to marginalize non-animal alternatives, advocates overestimate the human benefits attributable to vivisection and all but ignore the massive human harms that are an essential part of vivisection's legacy. Even more fundamentally, they uniformly fail to provide an intelligible methodology for comparing benefits and harms across species. I address each of these three failures in turn. (For a fuller critique, see my contribution to *The Animal Rights Debate*, Rowman & Littlefield, 2001).

CONCERNING THE OVERESTIMATION OF BENEFITS

Proponents of the Benefits Argument would have us believe that most of the truly important improvements in human health could not have been achieved without vivisection. The facts tell a different story. Public health scholars have shown that animal experimentation has made at best only a modest contribution to public health. By contrast, the vast majority of the most important health advances have resulted from improvements in living conditions (in sanitation, for example) and changes in personal hygiene and lifestyle, none of which has anything to do with animal experimentation. (For a summary of the relevant literature, see Hugh Lafollette and Niall Shanks, *Brute Science: Dilemmas of Animal Experimentation*, Routledge, 1996).

CONCERNING THE FAILURE TO ATTEND TO MASSIVE HUMAN HARMS

Advocates of the Benefits Argument conveniently ignore the hundreds of millions of deaths and the uncounted illnesses and disabilities that are attributable to reliance on the 'animal model' in research. Sometimes the harms result from what reliance on vivisection makes available; sometimes they result from what reliance on vivisection prevents. The deleterious effects of prescription medicines are an example of the former.

Prescription drugs are first tested extensively on animals before being made available to consumers. As is well known, there are problems involved in extrapolating results obtained from studies on animal beings to human beings. In particular, many medicines that are not toxic for test animals prove to be highly toxic for human beings. How toxic? It is estimated that 100,000 Americans die and some two million are hospitalized annually because of the harmful effects of the prescription drugs they are taking. That makes prescription drugs the fourth leading cause of death in America, behind only heart disease, cancer, and stroke, a fact that, without exception, goes unmentioned by the Benefits Argument's advocates.

Massive harm to humans also is attributable to what reliance on vivisection prevents. The role of cigarette smoking in the incidence of cancer is a case in point. As early as the 1950s, human epidemiological studies revealed a causal link between cigarette smoking and lung cancer. Nevertheless, repeated efforts, made over more than 50 years, rarely succeeded in inducing tobacco-related cancers in animals. Despite the alarm sounded by public health advocates, for decades governments around the world refused to mount an educational campaign to inform smokers about the grave risks they were running. Today, one in every five deaths in the United States is attributable to the effects of smoking, and fully 60 per cent of direct health care costs in the United States go to treating tobacco-related illnesses.

How much of this massive human harm could have been prevented if the results of vivisection had not (mis)directed government health care policy? It is not clear that anyone knows the answer beyond saying, 'A great deal. More than we will ever know.' One thing we do know, however: advocates of the Benefits Argument contravene the logic of their argument when they fail to include these harms in their defence of vivisection.

BENEFITS AND HARMS ACROSS SPECIES

Not to go unmentioned, finally, is the universal failure of vivisection's defenders to explain how we are to weigh benefits and harms across species. Before we can judge that vivisection's benefits for humans greatly exceed vivisection's harms to other animals, someone needs to explain how the relevant comparisons should be made. How much animal pain equals how much human relief from a drug that was tested on animals, for example? It does not suffice to say, to quote the American philosopher Carl Cohen (Cohen is the world's leading defender of the Benefits Argument) that 'the suffering of our species does seem somehow to be more important than the suffering of other species' (*The Animal Rights Debate*, op. cit., p. 291). Not only does this fail to explain how much more important our suffering is supposed to be, it offers no reason why anyone should think that it is.

Plainly, unless or until those who support the Benefits Argument offer an intelligible methodology for comparing benefits and harms across species, the claim that human benefits derived from vivisection greatly exceed the harms done to animals is more in the nature of unsupported ideology than demonstrated fact. (I note, parenthetically, that this challenge must be met by any contributor to this volume who uses this argument; if they fail to provide the necessary methodology, thoughtful readers will place no credence in what they say.)

HUMAN VIVISECTION AND HUMAN RIGHTS

The Benefits Argument suffers from an even more fundamental defect. Despite appearances to the contrary, the argument begs all the most important questions; in particular, it fails to address the role that moral rights play in assessing harmful, non-therapeutic research on animals. The best way to understand its failure in this regard is to position the argument against the backdrop of human vivisection and human rights.

Human beings have been used in harmful, non-therapeutic experiments for thousands of years. Not surprisingly, most human 'guinea pigs' have not come from the wealthy and educated, not from the dominant race, not from those with the power to assert and enforce their rights. No, most of human vivisection's victims have been coercively conscripted from the ranks of young children (especially orphans), the elderly, the severely developmentally disabled, the insane, the poor, the illiterate, members of 'inferior' races, homosexuals, military personnel, prisoners of war, and convicted criminals, for example. One such case will be considered below.

The scientific rationale behind vivisecting human beings needs little explanation. Using human subjects in research overcomes the difficulty of extrapolating results from another species to our species.

If 'benefits for humans' establishes the morality of animal vivisection, should we favour human vivisection instead? After all, research using members of our own species promises even greater benefits.

No serious advocate of human rights (and I count myself among this number) can support such research. This judgment is not capricious or arbitrary; it is a necessary consequence of the logic of basic moral rights, including our rights to bodily integrity and to life. This logic has two key components. (For a more complete discussion of rights, see my *The Case for Animal Rights*, University of California Press, 1983.)

First, possession of these rights confers a unique moral status. Those who possess these rights have a kind of protective moral shield, an invisible 'No Trespassing' sign, so to speak, that prohibits others from injuring their bodies, taking their life, or putting them at risk of serious harm, including death. When people violate our rights, when they 'trespass on our moral property', they do something wrong to us directly.

This does not mean that it must be wrong to hurt someone or even to take their life. When terrorists exceed their rights by violating ours, we act within our rights if we respond in ways that can cause serious harm to the violators. Still, what we are free to do when someone violates our rights does not translate into the freedom to override their rights without justifiable cause.

Second, the obligation to respect others' rights to bodily integrity and to life trumps any obligation we have to benefit others. Even if society in general would benefit if the rights of a few people were violated, that would not make violating their rights morally acceptable to any serious defender of human rights. The rights of the individual are not to be sacrificed in the name of promoting the general welfare. This is what it means to affirm our rights. It is also why the basic moral rights we possess, as the individuals we are, have the great moral importance that they do.

WHY THE BENEFITS ARGUMENT BEGS THE QUESTION

Once we understand why, given the logic of moral rights, respect for the rights of individuals takes priority over any obligation we might have to benefit others, we can understand why the Benefits Argument fails to justify vivisection on non-human animals. Clearly, all that the Benefits Argument *can* show is that vivisection on non-human animals benefits human beings. What this argument *cannot* show is that vivisecting animals for this purpose is morally justified. And it cannot show this because the benefits humans derive from vivisection are irrelevant to the question of animals' rights. We cannot show that animals have no right to life, for example, because we benefit from using them in experiments that take their life.

It will not suffice here for advocates of the Benefits Argument to insist that 'there are no alternatives' to vivisection that will yield as many human benefits. Not only is this reply more than a little disingenuous, since the greatest impediment to developing new scientifically valid non-animal alternatives, and to using those that already exist, is the hold that the ideology of vivisection currently has on medical researchers and those who fund them. In addition, this reply fails to address the substantive moral issues. *Whether* animals have rights is not a question that can be answered by saying how much vivisection benefits human beings. No matter how great the human benefits might be, the practice is morally wrong if animals have rights that vivisection violates. But *do* animals have any rights? The best way to answer this question is to begin with an actual case of human vivisection.

THE CHILDREN OF WILLOWBROOK

Willowbrook State Hospital was a mental hospital located in Staten Island, one of New York City's five boroughs. For 15 years, from 1956 to 1971, under the leadership of New York University Professor Saul Krugman, hospital staff conducted a series of viral hepatitis experiments on thousands of the hospital's severely retarded children, some as young as three years old. Among the research questions asked: Could injections of gamma globulin (a complex protein extracted from blood serum) produce long-term immunity to the hepatitis virus?

What better way to find the answer, Dr Krugman decided, than to separate the children in one of his experiments into two groups. In one group, children were fed the live hepatitis virus and given an injection of gamma globulin, which Dr Krugman believed would produce immunity; in the other group, children were fed the virus but received no injection. In both cases, the virus was obtained from the faeces of other Willowbrook children who suffered from the disease. Parents were asked to sign a release form that would permit their children to be 'given the benefit of this new preventive.' The results of the experiment were instrumental in leading Dr Krugman to conclude that hepatitis is not a single disease transmitted by a single virus; there are, he confirmed, at least two distinct viruses that transmit the disease, what today we know as hepatitis A and hepatitis B.

Everyone agrees that many people have benefited from this knowledge and the therapies Dr Krugman's research made possible. Some question the necessity of his research, citing the comparable findings that Baruch Blumberg made by analysing blood antigens in his laboratory, where no children were put at risk of grievous harm. But even if we assume that Dr Krugman's results could not have been achieved without experimenting on his uncomprehending subjects, what he did was wrong.

The purpose of his research, after all, was not to benefit each of the children. If that was his objective, he would not have withheld injections of gamma globulin from half of them. *Those* children certainly could not be counted among the intended beneficiaries. (Thus the misleading nature of the parental release form: not *all* the children were 'given the benefit of this new preventive'.) Instead, the purpose of the experiment was possibly to benefit some of the children (the ones who received the injections) and to gain information that would benefit other people in the future.

No serious advocate of human rights can accept the moral propriety of Dr Krugman's actions. By intentionally infecting all the children in his experiment, he put each of them at risk of serious harm. And by withholding the suspected means of preventing the disease from half the children, he violated their rights (not to mention those of their parents) twice over: first, by wilfully injuring their body; second, by risking their very life. This grievous breach of ethics finds no justification in the benefits others derived. To violate the moral rights of the few is never justified by adding the benefits for the many.

THE BASIS OF HUMAN RIGHTS

Those who deny that animals have rights frequently emphasize the uniqueness of human beings. We not only write poetry and compose symphonies, read history and solve math problems; we also understand our own mortality and make moral choices. Other animals do none of these things. That is why we have rights and they do not.

This way of thinking overlooks the fact that many human beings do not read history or solve math problems, do not understand their own mortality or make moral choices. The profoundly retarded children Dr Krugman used in his research are a case in point. If possession of the moral rights to bodily integrity and life depended on understanding one's mortality or making moral choices, for example, then those children lacked these rights. In their case, therefore, there would be no protective shield, no invisible 'No Trespassing' sign that limited what others were free to do to them. Lacking the protection these rights afford, *there would not have been anything about the moral status of the children themselves* that prohibited Dr Krugman from injuring their bodies, taking their lives, or putting them at risk of serious harm. Lacking this protection, Dr Krugman did not – indeed, he could not – have done anything wrong to the children. Again, this is not a position any serious advocate of human rights can accept.

But what is there about those of us reading these words, on the one hand, and the children of Willowbrook, on the other, that can help us understand how they can have the same rights we claim for ourselves? Where will we find the basis of our moral equality? Not in the ability to write poetry, make moral choices, and the like. Not in human biology, including facts about the genetic make-up humans share. All humans are (in some sense) biologically the same. But biological facts are indifferent to moral truths. Who has what genes has no moral relevance to who has what rights. Whatever else is in doubt, this we know.

But if not in some advanced cognitive capacity or genetic similarity, then where might we find the basis of our equality? Any plausible answer must begin with the obvious: the differences between the children of Willowbrook and those who read these words are many and varied. We do not denigrate these children when we say that our life has a richness that theirs lacked. Few among us would trade our life for theirs, even if we could.

Still, as important as these differences are, they should not obscure the similarities. For, like us, these children were the subjects of a life, *their* life, a life that was experientially better or worse for the child whose life it was. Like us, each child was a unique somebody, not a replaceable something. True, they lacked the ability to read and to make moral choices, for example. Nevertheless, what was done to these children, both what they experienced and what they were deprived of, mattered to them, as the individuals they were, just as what is done to us, when we are harmed, matters to us.

In this respect, as the subjects of a life, we and the children of Willowbrook are the same, are equal. Only in this case, our sameness, our equality is important morally. Logically, we cannot claim that harms done to us matter morally, but that harms done to these children do not. Relevantly similar cases should be judged similarly. This is among the first principles of rational thought, a principle that has immediate application here. Logically, we cannot claim our rights to bodily integrity and to life, and then deny these same rights in the case of the children. Without a doubt, the children of Willowbrook had rights, if we do.

WHY ANIMALS HAVE RIGHTS

We routinely divide the world into animals, vegetables, and minerals. Amoeba and paramecia are not vegetables or minerals; they are animals. No one engaged in the vivisection debate thinks that the use of such simple animals poses a vexing moral question. By contrast, everyone engaged in the debate recognizes that using non-human primates must be assessed morally. All parties to the debate, therefore, must 'draw a line' somewhere between the simplest forms of animate life and the most complex, a line that marks the boundary between those animals that do, and those that do not, matter morally. One way to avoid some of the controversies in this quarter is to follow Charles Darwin's lead. When he compares (these are his words) 'the Mental Powers of Man and the Lower Animals,' Darwin restricts his comparison to humans and non-human mammals.

His reasons for doing so depend in part on structural considerations. In all essential respects, these animals are physiologically like us, and we, like them. Now, in our case, an intact, functioning central nervous system is associated with our capacity for subjective experience. For example, injuries to our brain or spinal cord can diminish our sense of sight or touch, or impair our ability to feel pain or remember. By analogy, Darwin thinks it is reasonable to infer that the same is true of animals who are most physiologically similar to us. Because our central nervous system provides the physical basis for our subjective awareness of the world, and because the central nervous system of other mammals resembles ours in all the relevant respects, it is reasonable to believe that their central nervous system provides the physical basis for their subjective awareness.

Of course, if attributing subjective awareness to non-human mammals was at odds with the implications of evolutionary theory, or if this made their behaviour inexplicable, Darwin's position would need to be abandoned. But just the opposite is true. That these animals are subjectively present in the world, Darwin understands, is required by evolutionary theory. And far from making their behaviour inexplicable, their behaviour is parsimoniously explained by referring to their mental capacities.

For example, these animals enjoy some things and find others painful. Not surprisingly, they act accordingly, seeking to find the former and avoid the latter. Moreover, both humans and other mammals share a family of cognitive abilities (we both are able to learn from experience, remember the past, anticipate the future) as well as a variety of emotions (Darwin lists fear, jealousy, and sadness). Not surprisingly, again, these mental capacities play a role in how these animals behave. For example, other mammals will behave one way rather than another because they remember which ways of acting had pleasant outcomes in the past, or because they are afraid or sad. Concluding his comparison of the mental faculties of humans and 'the higher animals' (by which he means other mammals), Darwin writes: '[T]he difference in mind between man and the higher animals . . . is one of degree and not of kind' (*The Descent of Man*, Chapter IV).

The psychological complexity of mammals (henceforth 'animals,' unless otherwise indicated) plays an important role in arguing for their rights. Just as it is true in our case, so is it true in theirs: they are the subjects of a life, *their* life, a life that is experientially better or worse for the one whose life it is. Each is a unique somebody, not a replaceable something. True (like the children of Willowbrook), they lack the ability to read, write, or make moral choices. Nevertheless, what is done to animals, both what they experience and what they are deprived of, matters to them, as the individuals they are, just as what was to done to the children of Willowbrook, when they were harmed, mattered to them.

In this respect, as the subjects of a life, animals are our equals. And in this case, our sameness, our equality, is important morally. Logically, we cannot maintain that harms done to us matter morally, but that harms done to animals do not matter morally. Relevantly similar cases must be judged similarly. As was noted earlier, this is among the first principles of rational thought, and one that again has immediate application here. Logically, we cannot claim our rights to bodily integrity and life, or claim these same rights for the children of Willowbrook, then deny them when it comes to animals. Without a doubt, animals have rights, if humans do.

SOME OBJECTIONS, SOME REPLIES

Many are the objections raised against animal rights. While each is well intended, none withstands critical examination. It is to be recalled that the rights in question are the moral rights to bodily integrity and to life. Here, briefly, are the most important objections and my replies.

1 *Objection*: Animals do not understand what rights are. Therefore, they have no rights.
   Reply: The children of Willowbrook, all children for that matter, do not understand what rights are. Yet we do not deny rights in their case, for this reason. To be consistent, we cannot deny rights for animals, for this reason.
2 *Objection*: Animals do not respect our rights. For example, lions sometimes kill innocent people. Therefore, they have no rights.
   Reply: Children sometimes kill innocent people. Yet we do not deny rights in their case, for this reason. To be consistent, we cannot deny rights for animals, for this reason.
3 *Objection*: Animals do not respect the rights of other animals. For example, lions kill wildebeests. Therefore, they have no rights.
   Reply: Children do not always respect the rights of other children; sometimes they kill them. Yet we do not deny rights in their case, for this reason. To be consistent, we cannot deny rights for animals, for this reason.

4 *Objection*: If animals have rights, it follows that we will need to make arrangements for them to vote, marry, file for divorce, and immigrate, for example, which is absurd. Therefore, animals have no rights.

Reply: Yes, this is absurd; but these absurdities do not follow from claiming rights to life and bodily integrity in the case of animals any more than they follow from recognizing the rights of the children of Willowbrook, for example.

5 *Objection*: If animals have rights, then mosquitoes and roaches have rights. This would make it wrong to kill them, which is absurd. Therefore, animals have no rights.

Reply: Not all animals have rights because some animals do. In particular, neither mosquitoes nor roaches have the kind of physiological complexity associated with being a subject of a life. In their case, therefore, we have no good reason to believe that they have rights, even while we have abundantly good reason to believe that other animals (mammals in particular) do.

6 *Objection*: If animals have rights, then so do plants, which is absurd. Therefore, animals have no rights.

Reply: 'Plant rights' do not follow from animal rights. We have no reason to believe, and abundant reason to deny, that carrots and cabbages are subjects of a life. We have abundantly good reason to believe, and no good reason to deny, that mammals are. That is the morally relevant difference. In claiming rights for animals, therefore, we are not committed to claiming rights for plants.

7 *Objection*: Human beings are closer to us than animals are; we have special relations to them. Therefore, animals have no rights.

Reply: We do have special relations to humans that we do not have to other animals. We also have special relations to our family and friends that we do not have to other human beings. But we do not conclude that other humans do not have rights, for this reason. To be consistent, we cannot deny rights for animals, for this reason.

8 *Objection*: Only human beings live in a moral community where rights are understood. Therefore, all human beings, and only human beings, have rights.

Reply: At least among terrestrial forms of life, only human beings live in a moral community in which rights are understood. But it does not follow that only human beings have rights. It is also true that, at least among terrestrial forms of life, only human beings live in a scientific community in which genes are understood. But we do not conclude that therefore only human beings have genes. Neither should we conclude that only human beings have rights because only humans live in a moral community in which rights are understood.

9 *Objection*: Animals have some rights to bodily integrity and life, but the rights they have are not equal to human rights. Therefore, human vivisection is wrong, but animal vivisection is not.

Reply: This objection begs the question; it does not answer it. What morally relevant reason is there for thinking that humans have greater rights than animals? Certainly it cannot be any of the reasons examined in 1–8. But if not any of them, then what? The argument does not say.

The objections just reviewed have been considered because they are the most important, not because they are the least convincing. Their failure, individually and collectively, goes some way towards suggesting the logical inadequacy of the anti-animal rights position. Morality is not mathematics certainly. In morality, there are no proofs like those we find in geometry. What we can find, and what we must live with, are principles and values that have the best reasons, the best arguments on their side. The principles and values that pass this test, whether most people accept them or not, are the ones that should guide our lives. Given this reasonable standard, the principles and values of animal rights should guide our lives.

CONCLUSION

As was noted at the outset, animals are used in laboratories for three main purposes: education, product safety testing and experimentation (harmful non-therapeutic experimentation in particular). Of the three, the latter has been the object of special consideration. However, the implications for the remaining purposes should be obvious. Any time any animal's rights are violated in pursuit of benefits for others or, as in so-called 'basic research,' in pursuit of knowledge for its own sake, what is done is wrong. It is conceivable that some uses of animals for educational purposes (for example, having students observe the behaviour of rehabilitated animals when they are returned to their natural habitat) might be justified. By contrast, it is not conceivable that using animals in product testing can be. Harming animals to establish what is safe for humans is an exercise in power, not morality. In the moral universe, animals are not our tasters; we are not their kings.

The implications of animal rights for vivisection are both clear and uncompromising. Vivisection is morally wrong. It should never have begun and, like all great evils, it ought to end, the sooner the better. To reply (again) that 'there are no alternatives' not only misses the point, it is false. It misses the point because it assumes that the benefits humans derive from vivisection are derived morally when they are not. And it is false because, apart from using already existing and developing new non-animal research techniques, there is another, more fundamental alternative to vivisection. That is to stop doing it. When all is said and done, the only adequate moral response to vivisection is empty cages, not larger cages.

Essay Three

WHY ANIMALS' RIGHTS ARE WRONG
Stuart Derbyshire

My position on animal research in medicine is uncompromisingly human centric. The attempt to draw comparison between humans and animals is fundamentally problematic. Human beings have agency and an appreciation of their circumstances that places them outside of the natural dictates that order and regulate the lives of animals. The animal world is a constituent part of nature but our world is not. These facts are important because they render the proposed extension of rights to animals nonsensical. Animals are not beings of a kind that can exercise or respond to any type of claim against them. As Carl Cohen, Professor of Philosophy at the University of Michigan, has argued, animals and humans occupy separate moral spheres that cannot be forced into compliance (see, for example, his article 'The case for the use of animals in biomedical research', *New England Journal of Medicine* 315: 865–70, 1986).

Advocating animal rights involves a certain exaggeration of animal capacity but also relies upon a denigration of human ability that is much more dangerous. Our ability to understand nature and bend it towards our needs takes on an organized form through the project of science. Science occupies a special realm within the gamut of human behaviour because it increases the possibility for human action and therefore advances the cause of human freedom. When science is upbraided for intruding upon nature it is not only science but also humanity that is being attacked.

The stakes in this debate are of the highest order, which in part explains the excitement and intensity that the argument can inspire. The sometimes dangerous and often hostile response to animal research has driven many scientists to be defensive. Animal experiments are increasingly performed in a state of isolation behind hi-tech security walls. While perhaps understandable, the development of bunker research is counterproductive because it reinforces the view of scientists as partaking in suspicious, dangerous and immoral activities. Defensive in their behaviour, scientists have also become defensive in their arguments, making efforts to accommodate to the animal rights movement, or to reason with it and make compromises. These efforts are also counterproductive, reinforcing the legitimacy of animal rights campaigns and suggesting the illegitimacy of animal research.

THE PROBLEM OF THE THREE Rs

The most widespread accommodation to animal rights is the adoption of the three Rs, first proposed in 1959 following a report for The Universities Federation for Animal Welfare (UFAW). The three Rs are 'refinement', 'reduction' and 'replacement'. Scientists pledge to refine their techniques so as to induce the minimum amount of suffering; reduce the number of animals used; and replace animals with other techniques wherever possible.

At first blush the three Rs appear reasonable, if somewhat patronizing. All animal experimenters know to reduce the amount of stress an animal is subjected to (refinement) so as to not hinder discovery – a stressed animal will be less likely to behave or respond normally. Equally, all researchers will naturally tend to use fewer or less-costly animals or techniques (reduction and replacement) so as to get quicker results for fewer resources.

Patronizing or not, the three Rs were not developed from the perspective of good scientific practice. They were developed from the perspective of animal welfare. This makes the three Rs disastrous, reinforcing a low-life opinion of animal researchers and encouraging the notion that animal experiments are problematic. Once the perspective of the animal is adopted, it is inevitable that all experimentation will be seen negatively, as no animal experiments are in the interests of the animal. Adoption of the three Rs comes across as a confession of guilt. The impression is that research animals are a 'necessary evil', when, in fact, they are necessary, period.

THE NECESSITY OF ANIMAL RESEARCH

The transplant pioneer at the University of Pittsburgh, Dr Thomas Starzl, was once asked why he used dogs in his work (see K. Goodwin and A.R. Morrison, 'Science and Self Doubt', *Reason* October, 2000). He explained that his first series of kidney transplant operations left the majority of his subjects dead. He figured out what enabled the minority to survive and commenced a second series of operations; the majority of these subjects lived. A third group of subjects received liver transplants and only one or two died. In his fourth group all subjects survived. Starzl remarked that it is important to realize that his first three groups of subjects were dogs, while the fourth group was human babies. Was he supposed to experiment and refine his technique on humans, or was he expected to abandon a promising line of research that has saved innumerable lives? It is remarkable that we even have to consider the question.

The example of Starzl's research dramatically illustrates the importance of animal research in developing medicine. There are many other examples. In the 1950s, primate researchers developed chlorpromazine, used to treat mental illnesses such as schizophrenia. In the 1960s, monkeys were used to develop a vaccine against rubella and in the surgical transplanting of corneas to restore vision. In the 1970s and 1980s, primate research helped track down tumour viruses, to improve chemotherapy. The now widely used vaccine for hepatitis B was developed largely in chimpanzees. The current vaccine candidates against AIDS were all developed using primates. Researchers learned to do organ transplants using macaque monkeys. Potent anti-rejection drugs, such as Cyclosporin, were first used in non-human primates. The design of the heart-lung transplant was developed in rhesus macaques. Control of diphtheria came from guinea pigs and horses. Open-heart surgery was developed in dogs, as was the technique of kidney transplantation. Critical diabetes work, leading to the development of insulin, was carried out in dogs. From sheep came control of anthrax, and from cows the eradication of smallpox.

There are considerably more breakthroughs that can be attributed, wholly or in part, to the use of animal experimentation, including the eradication of polio and the control of rhesus incompatibility (see S.W.G. Derbyshire, 'Animal research: a scientist's defence', Spiked, 2001 for further comments and details (www.spiked-online.com)). There is dissent regarding the overall value of animal research (J. B. McKinlay and S. McKinlay, Health and Society, Milbank, 1977) and considerable argument over particular famous events such as the development of penicillin and the thalidomide tragedy. A fair assessment of the history of animal research is that animal work has provided considerable breakthroughs in knowledge and understanding while at the same time, and consistent with all scientific endeavours, there have been many hopeless experimental failures and the occasional calamity.

Those who argue against the continued use of animals in research and other endeavours rarely do so by suggesting that humanity has not gained from the use of animals, which is a difficult argument to make. Instead it is proposed that animals retain capacities that make our use of them, as a means to an end, immoral. One advocate of this position is Tom Regan, Professor of Philosophy at North Carolina State University. Regan argues that if animals have a capacity for suffering, feeling and thinking as humans do then it should logically be the case that they have a claim to rights similar to ourselves. Anything else would be 'speciest' (a term first coined by Richard Ryder, see his *Victims of Science*, Davis-Poynter 1975). Regan is an abolitionist; he wishes to prevent all forms of animal use for human ends. That such an extreme view remains influential is testament, in part, to Regan's doubtless ability in constructing a challenging argument. For this reason it is worth spending time to consider what Regan has to say. The influence of abolitionist views, however, is also testament to the deep ambivalence that our society now has towards science and the development of knowledge. Being 'pro-animals' is a natural adjunct to the anti-humanism reflected in our ambivalence towards the project of human advancement.

THE INTUITIVIST STANCE AND THE CASE FOR ANIMAL RIGHTS

Regan argues from a point of principle, which is that animals have cognitions and awareness comparable to our own and should, therefore, be treated as having 'inherent value'. The success or failure of animal research is beside the point; the value of animal life is sufficient that we should not use them as a means to an end. We do not use people, even mentally disabled persons, no matter how extreme their disability, as a means to an end and so, intuitively, if animals have equal abilities to persons of varying mental capacity, according to Regan, there is equal reason to defend the value of the animal as much as people. In case I have inflicted violence on Regan's stance, let me quote him directly:

> My position, roughly speaking, may be summarized as follows. Some non-human animals resemble normal humans in that, like us, they bring the mystery of a unified psychological presence to the world. Like us, they possess a variety of sensory, cognitive, conative, and volitional capacities. They see and hear, believe and desire, remember and anticipate, and plan and intend. Moreover, as is true in our case, what happens to them matters to them. Physical pleasure and pain – these they share with us. But they also share fear and contentment, anger and loneliness, frustration and satisfaction, and cunning and imprudence; these and a host of other psychological states and dispositions collectively help define the mental lives and relative well-being of those humans and animals who (in my terminology) are 'subjects of a life'
>
> (T. Regan, *Defending Animal Rights*, University of Illinois Press, 2001: 42–43).

Ryder also flags the mental abilities of animals, especially the capacity for experiencing pain, as a driving force of animal rights advocacy. Ryder has coined the term 'painience' to describe the capacity to suffer pain or distress of any sort and makes this the basis for rights qualification (R. Ryder, 'Darwinism, altruism and painience', A talk presented to Animals, People and the Environment, 1999 (www.ivu.org/ape/talks/ryder/ryder.htm)). Cohen agrees that animals experience pain but does not view this as guaranteeing animals access to rights but rather as enforcing the duty of humans not to be cruel (C. Cohen, T. Regan, *The Animal Rights Debate*. Rowman and Littlefield Publishers, 2001).

In my opinion, Regan, Ryder and Cohen are all guilty of exaggerating the capacities of animals. They draw false equivalence between the capacities and abilities of humans and animals based on a superficial contrast leading to the moral blunder of equating the inequitable. I also feel they are guilty of minimizing the capacities of humans. This is especially true of Ryder who views the activities of humanity as reducible to the avoidance of pain like some kind of automatons which reflexively respond more to the good stuff and less to the bad.

THE QUESTION IS, 'CAN THEY SUFFER?'

The famous dictum of the philosopher Jeremy Bentham, 'The question is not, Can they *reason*? nor, Can they *talk*? but, Can they *suffer*?' is routinely wielded in defence of animal interests with the assumed positive answer to the latter question. But this assumption is hasty. There is good evidence that the extent to which animals suffer and feel pain is minimal, especially when compared to our own. The suggested experience of pain in animals is an interpretation based on our own experience that we project onto the animal world. The projection is understandable but wrong, as an examination of the term 'pain' will reveal (see S.W.G. Derbyshire, 'The IASP definition captures the essence of pain experience', *Pain Forum* 1999; 8:106–9; S. W. G. Derbyshire, 'Locating the beginnings of pain', *Bioethics* 1999; 13:1–31; S.W.G. Derbyshire, 'Fetal pain: An infantile debate', *Bioethics* 2001; 1:77–84, for further discussion).

To avoid the tautological use of the term pain we need some sort of definition of what pain is. In the absence of a definition, pain is ascribed when there are 'pain behaviours' or 'pain stimuli' or, more simply, pain is present when there is pain. The circularity arises because the description of behaviours that follow noxious insult and the psychological description of pain experience are at two different levels: the link between them is at the heart of the problem of understanding animal suffering or other types of animal experience (see T. Nagel, 'What is it like to be a bat?', *Philosophical Review* 1974; 4:435–50). That link, however, is rarely explored or even recognized and is instead simply assumed.

To escape the inherent tautology of describing pain as the result of painful stimuli, pain is generally defined as a sort of amalgam of cognition, sensation and affective processes, commonly described under the rubric of the 'biopsychosocial' model of pain (discussed in S.W.G. Derbyshire, 'Sources of variation in assessing male and female responses to pain', *New Ideas in Psychology* 1997; 15:83–95 and G. Waddell, 'A new clinical model for the treatment of low-back pain', *Spine* 1987; 12:632–44). Pain is no longer regarded as merely a physical sensation of noxious stimulus and disease, but is seen as a conscious experience that includes mental, emotional and sensory mechanisms. Pain has been described as a multidimensional phenomena for some time and this understanding is reflected in the current IASP (International Association for the Study of Pain) definition of pain as 'an unpleasant sensory and emotional experience associated with actual or potential tissue damage, or described in terms of such damage' (H. Merskey, 'The definition of pain', *European Journal of Psychiatry* 1991; 6:153–9). Although clearly not without its flaws, the IASP definition captures the basic essence of pain subjectivity and I have defended its continued use elsewhere.

The common-sense view of pain as a low-level phenomenon directly contingent upon injury is, by this definition, a mistaken one. Pain is actually a high-level process that makes no sense in the absence of sentience; pain accompanies injury in minds that are capable of subjectivity and this criterion is a reach for animals. There are good reasons for believing that animals lack the ability for reflection (and therefore lack an inner world) and the capacity for reasoning or suffering.

In his work, *Kinds of Minds*, Dennett (1996) muses over the sorts of thinking a talking lion may have:

> ... if a lion could talk, we could understand him just fine – with the usual sorts of effort required for translation between different languages – but our conversations with him would tell us next to nothing about the minds of ordinary lions, since his language-equipped mind would be so different. It *might* be that adding language to a lion's 'mind' would be *giving* him a mind for the first time! Or it might not. In either case, we should investigate the prospect and not just assume, with tradition, that the minds of non-speaking animals are really like ours.
>
> (D. C. Dennett, *Kinds of Minds: Towards an Understanding of Consciousness*,
> Basic Books, 1996: p.18, emphasis in the original)

Dennett raises a number of issues that are pertinent to the discussion at hand. Given the nature of the lion, what kind of 'mind' is it likely to have? If you add language, or other cognitive constructs, how might that alter the mind? Dennett warns that conceptualizing the problem of an animal's 'mind' can easily become a tautology that assumes precisely what it is that needs to be explained. If something thinks then it must think particular thoughts. Particular thoughts, however, are composed of particular concepts. 'Ouch', is not just a reflexive response, but is a mindful state that includes the sensation and the associated cognitions and emotion. How might we express, in whatever form, the precise 'experience' of pain (or 'ouch') that the lion or other animal is proposed to be experiencing?

To probe the answer to this it is worth considering how we humans come to have subjectivity. The contents of our inner world comes to mean something to us only in so far as it means something to others and the realization of that meaning occurs through a developmental process that is social as well as natural. Pain is a good example, being a clearly subjective, personal, phenomenon but not one that is locked behind a steel door. If pain were an entirely private affair, no words would be able to express it because no external frame of reference would be comparable and therefore adequate to express the sensation. Pain is not like this because clearly human beings do express their painful experiences and these expressions have meaning that allow for diagnosis, treatment and eradication of pain. In so far as human beings live in a community of thinking, feeling, talking beings the privacy of experience is broken down and externalized for further analysis. As we are able to externalize our inner world, so we are able to reflect upon that world and become self-aware or self-conscious. Consciousness *is* self-consciousness, one cannot reflect upon the world without knowing that it is I who am reflecting. If we were not conscious of being conscious, then we would be unconscious of consciousness, which is an absurdity.

The evidence for this type of self-awareness within the animal world is limited and controversial (see, C. M. Heyes, 'Anecdotes, trapping and triangulating: do animals attribute mental states?', *Animal Behavior* 1993; 46:177–88; K. Malik, *Man, Beast and Zombie*, Weidenfeld, 2000; J. Vauclair, 'Mental states in animals: cognitive ethology', *Trends in Cognitive Sciences* 1997; 1:35–9). There is some evidence of potentially reflective consciousness from behaviours such as giving of alarm calls and the use of deception. Taking the most charitable view and assuming these behaviours do indicate some internal recognition, there is good reason to believe that the animal experience, if there is any at all, is heavily circumscribed and does not generalize particularly far. There is no evidence that animal awareness, however defined or understood, provides any transformative impact within an individual animal or group. A chimpanzee today behaves in much the same way as a chimpanzee did 100,000 years ago. When chimps forage for food they do not ask themselves why or consider better alternatives any more than a beaver considers better ways of building dams. When swallows fly south in the winter they do not ask why it is hotter in Africa or what would happen if they flew further south or whether they could save themselves the bother by creating warmth in the north. Humans do ask these kinds of questions and do engage in behaviour that transforms their circumstances. We are not trapped inside a purely personal, solipsistic view of the world. We have insight.

Does this mean animals are absent pain and suffering? I believe that an honest appraisal of the evidence would suggest a largely 'yes' response. The behaviour we observe following a noxious insult to an animal is disturbing to us because we have insight and we project our expectation of feeling onto the animal. But behaviour can be misleading. For example, reflex responses to skin damage and other noxious insult can occur despite the absence of any higher cortical brain centres (those regions of the brain thought to be responsible for reasoning, language and the organization of behaviour) that are surely necessary to translate experience. Those reflex responses can look to us like behaviours that betray the inner feeling of pain. Our subjectivity has the promiscuous tendency to over-generalize. In contrast to ourselves, animal behaviour is mechanical, driven by the dictates of nature and immune to the processes of reflective cognition that we take for granted. It is a black, silent existence that is not conscious of its own processes or, *at the very most*, a dark murky experience that does not compare with our own.

ANIMAL RIGHTS AND WRONGS

The exaggeration of animal capacities occurs, in part, to enable the case for animal rights to be made by analogy. If animals can be made more like us, and us more like animals, then the case for animal rights can be made simply by pointing to our own rights that we tend to take for granted. A related tactic is to use the circumstances of those unfortunate humans who clearly lack certain capacities in order to push the case for animal rights. There are members of our own species who are either born with very limited mental capacity or descend into limited capacity through Alzheimer's disease and similar calamities. We do not deny these unfortunate individuals access to our moral world so why should we exclude the animals?

Peter Singer, Professor of Philosophy at Princeton University, and Tom Regan have adopted the term 'speciesism' that was introduced by Ryder as shorthand for excluding animals from this human sphere of moral consideration. Singer has argued that, in the same way that racists and sexists give greater weight to members of their own race or gender to advance their interests, so does a speciest give greater weight to their own species. According to Singer and Regan, we have no truck with racists or sexists and neither should we with speciests. Cohen is appalled:

> This argument is worse than unsound; it is atrocious. It draws an offensive moral conclusion from a deliberately devised verbal parallelism that is utterly specious.
>
> (C. Cohen, 'The case for the use of animals in biomedical research', *New England Journal of Medicine* 1986; 315:867)

The offence here is greater than that caused when merely suggesting animals could or should be included within our moral sphere. There is an implied equivalence between the campaign for animal rights and the struggle of blacks and other groups against oppression, which is insulting because it belittles those struggles. Animals care not one jot about rights and will not fight for them ever. The rub is that at least some mentally disabled persons care equally and are just as apathetic. Cohen's outrage is not misplaced, the parallel with the struggles against oppression by women and people of colour are a disgrace and an insult, but the anger cannot carry into the realm of the disabled. Something is clearly amiss here and resolution lies with a better understanding of what rights are.

DIFFERENT KINDS OF RIGHTS

Over the course of human history, various groups have had to fight for rights and equality. Not just against the self-interest of those with power but also against the very conditions of existence. It might appear to us today that rights are 'self-evident' but in fact they are dependent upon a certain level of production. Not to put too fine a point on it, when food was scarce, tyranny and slavery were inevitable. Freedom and independence for all became possible only as society was able to produce surplus.

Today the surplus production is sufficient that no group can be legitimately denied their freedom. To the extent that oppression exists today, it follows that a vested interest will have to be forcibly removed. It can be argued, and this might be the direction of Regan's thinking, that society is now sufficiently advanced that rights can and should be extended to non-human species. Whatever mental capacities animals may have, however, taking even the most charitable view, animals clearly lack the ability to recognize or take advantage of their historic opportunity. From the animals' perspective there is no opportunity; it simply does not exist. For this reason, Cohen has argued that animals are not and cannot be a part of the moral community. Cohen views animals as existing in a separate world that does not have moral content; animals are totally amoral, there is no morality for them, and they do no wrong, ever. In their world there are no rights. Animals are denied access to our moral world because they lack the capacity to take individual responsibility and cannot be held to account for their actions. Humans, on the other hand, are responsible for their actions and are therefore moral agents. Although it might appear that the extension of rights to animals is possible it is in fact impossible, the appearance is illusory.

The right to have freedom of thought and action independent of interference from the state or other body cannot be extended to animals because animals cannot exercise these rights in any meaningful way. These 'negative' rights are also denied children and those with certain levels of mental handicap for the same reasons. Children and those with certain handicaps are understood to lack the competence necessary to exercise freedom of thought and independence of action. In this regard it is degrading to the concept of freedom to confuse the protection of children and care of the handicapped with the concept of rights. Mental ability really does act as a screen for sorting those who will and those who will not have access to negative rights. The precise ability necessary is highly debatable but as a rule of thumb it does not seem unreasonable to exclude those who lack the capacity to recognize or care that the opportunity for freedom is there and/or to exclude those who lack the ability to survive upon receiving those freedoms. Animals clearly fail the former test; children and the mentally disabled may fail in the latter or both.

Regan argues that we should treat animals as if carrying invisible 'no trespassing' signs that ensure certain rights, such as the right not to be injured. Such 'positive' rights are in reality forms of protection rather than forms of freedom. Demands for the right not to be inconvenienced by cigarette smoke, loud noise, traffic and so on are contemporary examples of demands for protection masquerading as demands for rights. Unlike negative rights, these positive rights are difficult to universalize, the right not to breathe second-hand smoke, for example, bumps up against the rights of others to smoke. This is why positive rights are not really rights and are better understood as protections or privileges. The interests of one group are seen to be greater than that of another and are therefore given special privileges relative to the other group. We provide certain protections for children until they are of an age when they can care for themselves; we provide protection for the disabled to the extent that they cannot care for themselves. Unlike negative rights that generally free the individual from outside interference these protections are dependent upon external bodies whether that be the family, a nursing home, a park ranger or the courts and other branches of the state. The 'freedoms' that are provided by such protections, whilst important, cannot be correctly termed freedoms since they are dependent upon the good sense or good will of somebody else or some other group.

These protections are extensions of welfare provided to children, the disabled and, potentially, animals, and extensions of authority in those cases where the state steps in. Understood in this way, Regan's demand for animal rights is indistinguishable from a radical welfarist position. Both demand that animals be protected from animal research, farming and other uses for human ends.

THE QUESTION OF ANIMAL WELFARE

We can and do extend protection and welfare to animals but we do this in an arbitrary manner. Regan demands consistency. His trump card in this respect is, again, the rights of the disabled, or the protection of their welfare, as I would prefer to see it. Regan argues that we do not use humans in experiments, no matter what their state of disability may be. Whatever level of existence animals may possess we can imagine a human being in the same state and we would not allow experimentation on the human. Intuitively, therefore, it seems wrong for us to be inconsistent, how can we judge the moral value of a human being who is technically dead, alive only through the action of a respirator, for example, as being greater than that of a chimpanzee?

The main contrast that leads to the greater worth of our technically deceased human are the outside interests of society in general and the family in particular. When a human being is lost, the loss is felt at a social level. The potential that the human being represented to be productive, insightful and to provide a contribution passes with death and we mourn that loss. The loss is, of course, particularly acute for family and close friends who would have had first-hand experience of the actuality of the person's existence and hopes and aspirations for the potential of the deceased. We do violence to the value a human being represented or could have represented if we treat a human instrumentally, even in death.

In contrast, animals never have any potential to do anything greater than their ancestors and direct contemporaries. Animals are not individual because, while they may have distinct characteristics, they lack the capacity to develop themselves and transform their existence. Animals are also not social because while they may live within groups, they lack the capacity to transform that group's behaviour and they cannot take collective decisions within the group. In this sense, the value of animals is fixed such that it is always comparable to any other animal currently living, dead or projected into the future. When an animal dies, unless we have some particular association with the animal such as a pet, we do not mourn the passing because there is nothing to mourn. Animals never have the value that humans retain even when deceased, unless we provide some value through a human relation.

The value that animals have is derived from the relationship we provide for them. Pets occupy a special category and are protected within that category. There is nothing about the actual or potential existence of animals, however, that demands we provide those protections comprehensively. Our decisions are human-centric, arbitrary and consistent with the uses to which animals can be put by us. Animals do not possess 'inherent value' as Regan would have us believe.

None of this means that we can be destructive without purpose when it comes to animals or that animals should be treated without care or concern. Mindless destruction of any kind is rightly considered anti-social and so it should be with animals. Furthermore, whatever experience animals may or may not have, they certainly behave as if in discomfort and deliberate extension of such behaviour without reason is degrading. Where there is purpose, however, then the treatment of animals should be dictated according to the better living, knowledge and understanding of humanity.

WELFARE IN ANIMAL EXPERIMENTATION

From the perspective of animal research, which is surely the most purposeful of human activity with animals, the demand for animal welfare is simply ludicrous and threatens the integrity of scientific work. Concessions to animal welfare by animal researchers make no sense and are, in any case, unnecessary in that all scientists should strive to perform the best possible experiment. Such practice inevitably demands that animals are well looked after so as to be good test subjects. It makes no sense for animal researchers to engage in a discussion of animal welfare beyond ensuring that the animals will be properly housed, fed and exercised, and that they will be generally physically and behaviourally nourished as much as possible to benefit their performance as an experimental subject. The idea that we should – or even can – be any more concerned about their welfare stretches credibility.

Giving AIDS and other diseases to animals, carrying out experimental surgeries and infusing untested drugs hardly sound like procedures aimed at protecting the animals' welfare. Mistreating animals is unacceptable because it ruins experiments – but any further concern for animals' well-being is beside the point. Animal researchers and their advocates cannot have it both ways. Professed concern for the welfare of laboratory animals is simply inconsistent with the reality of laboratory experiments that almost invariably result in distress and death for the animal.

CONCLUSION

Suggestions that rights should be extended to animals are based on two misunderstandings, the first being a false equivalence between human and animal experience, and the second being a confused interpretation of rights. Behaviour is not a faithful reflection of experience and protection is a privilege not a right. The consequences of these misunderstandings are unfortunate to say the least. Considerable regulation micromanages animal research and dictates the direction of that research in ways that would be unthinkable for other branches of science. In the UK, for example, the researcher has to obtain a licence from the Home Office, a process that follows the principles of the three Rs. Invariably, considerable justifications for any procedures that involve distressing the animal will be required, and considerable pressure for the use of fewer animals, from further down the phylogenetic tree (such as using rats rather than primates), will be applied. In the USA, all universities have their own Institutional Animal Care and Use Committees (IACUCs), which voluntarily assess any proposed research with respect to the three Rs and provide assurance to grant awarding bodies that the research is animal-welfare friendly. Further regulation is possible following Congressional approval of an agricultural spending bill enabling the US Department of Agriculture to regulate the use of mice, rats and birds in scientific research. Currently the Association of American Medical Colleges (AAMC), the National Association for Biomedical Research (NABR) and the Federation of American Societies for Experimental Biology (FASEB) are disputing the need for further regulation. As we go to press it remains unclear which side is likely to prevail over the long haul.

Certainly the AAMC, NABR and FASEB face a difficult battle. We live in an age of deep ambivalence towards human progress with scientists bemoaned for 'playing God', 'interfering with nature' and creating destruction. War, nuclear annihilation, the holocaust and the many other failures and calamities of the past century battered people's trust in science and bruised their hopes for progress. The tendency to see humans as being more brutish, and animals as more human, is understandable but profoundly wrong and deeply troubling. Such anti-human prejudice can only further erode our confidence in the human project and threatens to condemn us to further darkness and superstition in the future.

Essay Four

INSTITUTIONAL SPECIESISM: CRUELTY IS WRONG

Richard D. Ryder

'Of course you must be experimented upon,' said the alien, 'I am vastly more intelligent than you are, so I am permitted to cause you pain. What is more, you are from a different species than mine.'

The frightened girl looked up at the monster hovering over her.

'But', she pleaded, 'what about my rights?'

'Ha!' replied the alien. 'Humans can have no rights because you have so little in the way of thought, autonomy and self-awareness compared with us.' His three eyes flashed proudly. 'Besides, we need to experiment upon you in the name of science ... and science is sacred.'

I remember, years ago, visiting a psychology laboratory in California. An American scientific colleague showed me around. There were the usual rats in cages and monkeys with electrodes cemented into their brains. Then we came upon the most terrible sight: a cage in the shape of a wheel was being slowly rotated by an electric motor. Inside it, flopping around and moaning, were the mangled remains of a white cat. It had been blinded and its tail had been amputated. 'Whatever is going on?' I asked. 'The wheel rotates continuously day and night for about a week,' replied the professor, 'it's a sleep deprivation experiment'.

As I watched, two students stopped the rotation, wrapped the cat in a blanket, held it firmly and plunged a needle into the base of its skull. The cat's moan turned to a scream. 'They're only second year students,' said the professor casually, 'and they've got some crazy idea that the combination of sleep deprivation and blinding will show up in the cerebrospinal fluid. They take samples every few hours.' Mistaking my revulsion for scientific scepticism, he added, 'Well, at least it gives them experience in handling animals.'

That was a turning point for me. Already deeply disturbed by what I had seen in other psychology laboratories in Cambridge, Edinburgh and New York, I decided that this was a career I could not stomach. Initially, I wondered if there was something wrong with me; whether I was abnormally sensitive, morally warped or (horror of horrors) lacking in 'masculinity'. Nobody else seemed to think that there was anything wrong with the torture I saw around me. There was little or no public criticism of vivisection at that time. It was only some five years later, in 1969 when I was working in a hospital in Oxford, that I summoned up the courage to speak out by writing letters to the newspapers. My book Victims of Science, criticizing research practices, was published in 1975 and, luckily, precipitated a lot of interest in Parliament and the media. did not ease up my peaceful campaigning until 1986 when, eventually the Animal (Scientific Procedures) Act was passed in Britain and legislation was introduced into the European Union in the same year.

SPECIESISM

What, then, are my objections to painful and distressing experiments on animals? First and foremost my objection is a moral one. Why should our species imagine that we have a right to treat other species in such a manner? What is so special about the human species that it can treat other animals in a way that we would not treat our worst enemies? What can possibly justify such speciesism? Like the imagined alien from outer space we tend to give ourselves outrageous excuses. Admittedly, we are (with exceptions) more intelligent than other animals but how does this justify our tyrannical behaviour? Does this mean that less intelligent humans (and, as a psychologist I have encountered professionally some severely handicapped people who were far less intelligent than most apes, dogs or cats) can be given lesser rights than the more intelligent? What does intelligence have to do with the possession of moral status? The answer is that it does not have any logical connection at all.

Other excuses have been used by vivisectors to try to justify their oppression and exploitation of animals. Non-human animals are said to lack language, technology or self-awareness. But, surely, all these qualities are morally irrelevant. Besides, we know that some animals, such as apes, can learn to use our sign languages. We may not be clever enough to learn their languages but they are clever enough to learn ours. Some primates fashion and use simple tools to access food. They also show self-awareness in that they can recognize their own images in a mirror. But even if they did none of these things it would not be significant in moral terms. All that matters, surely, is that other species can suffer pain, fear and distress in very much the same way that humans do. To discriminate against others merely because they have a different physical appearance is very unintelligent. Such speciesism is as irrational as sexism or racism.

PAINISM

It is our common painience that we share with other animals that puts us all into the same moral category. The suffering of pain or distress is what matters morally. Furthermore, it makes no difference who suffers it. This was pointed out years ago, in the eighteenth century, by the Reverend Humphry Primatt who wrote in 1776, 'Pain is pain, whether it be inflicted on man or beast'. The philosopher, Jeremy Bentham, made a similar point in 1789 when, in referring to the moral status of animals, he wrote, 'The question is not can they *reason*? Nor, can they *talk*? But can they *suffer*?'

The suffering of pain and distress has become the central issue in ethics today. It has taken time for other concepts – sanctity, reason, virtue, for example – to be put into their correct perspectives and for philosophers to realize that conditions such as justice, equality and liberty are all morally subordinate to the reduction of pain. Indeed, justice, equality and liberty are psychologically and ethically important precisely because they tend to have an analgesic effect. I have outlined this ethical position, which I call painism, in my book *Painism: A Modern Morality* (Opengate Press, 2001). Broadly, my conclusion is that our moral duty is to try to alleviate the sufferings of others equally, regardless of their species, race, gender, religion or ethnicity, and that we should give priority in this undertaking to the sufferings of those individuals who are suffering most severely.

Painism, as an ethical theory, applies, of course, to humans as well as to non-humans. It differs from some other ethical theories, such as Utilitarianism, in that I reject the practice of aggregating the pains and pleasures of *several* individuals when making a moral judgement. I believe it is impossible to add up the pains or pleasures of different individuals and then argue that the total score means anything. Each individual is his or her own experiential universe. I can never feel exactly the same pain that is any other individual's. The boundaries of consciousness are the boundaries of the individual.

DEBATING MATTERS

One huge problem with Utilitarianism is that it can justify cruelty on the grounds that the *aggregated* benefits to several others outweigh the pain of the victim. So, according to Utilitarianism, the pain of a gang rape victim or of a laboratory animal, for example, can be justified if the aggregated *total* of the consequent pleasures or benefits of all those affected by the action (for example, the rapists) is greater than the individual victim's pain. This is clearly absurd.

I do, however, accept that it is possible to justify the infliction of some pain on an individual for the greater benefit of *another single individual* – for example, restraining a bully from attacking a victim. (Please note that I use the word 'pain' to cover sufferings of any sort.)

THE INVALIDITY OF VIVISECTION

The body of a live animal (whether human or non-human) is rather like a complex machine. Its essential difference, of course, is that it is conscious and can suffer. However, imagine oneself to be a visiting alien from another planet encountering on Earth a motor car for the first time. How would we try to discover how it works, all text books on the subject having been destroyed? A certain amount could be worked out by stopping the car and taking it apart, piece by piece. But such 'post mortem' analysis would not supply all the answers. It would not provide a full picture of *processes*, such as motion, combustion, transmission and the interaction of the parts. These could more easily be understood while being observed in action, while the mechanism is, as it were, 'alive'.

This is the rationale for experimenting upon living animals. It has some merit although this is constantly being overstated. Indeed, if I look back on the recent history of my old field – experimental psychology – I can see that hundreds of thousands of animals have been used over the last hundred years, often being subjected to hideous cruelty such as blinding, deafening, electric shock, paralysis, terror or starvation. Almost every one of these experiments seemed, at the time, to be crucially important to the experimenter. Yet, at the end of the day, what has been learned? Precious little. Experimental psychology is not a medical subject. Indeed, from about 1970 the whole business of behavioural research on animals, as it had been practised since about 1900, started to be dismantled. After the reckless sacrifice of thousands of animal lives, even the devotees of strict Behaviourism had to admit that it was getting nowhere. In consequence, the field was all but abandoned. So much suffering had been caused and for no significant benefit.

As a psychology student at Cambridge, I had been taught that any results obtained from research carried out on non-human animals could *not* be applied to humans. It was considered almost a crime, in scientific terms, to try to apply results from one species to any other species. So, after about ten years of campaigning against experimentation on animals, it amused me to hear the scientific community reversing all these arguments. Scientists experimenting upon animals suddenly started assuring the public that their experiments were, of course, all essential for human benefit!

SCIENCE AS RELIGION

Science can take on many of the attributes of religion. Its worshippers believe fervently (even arrogantly) in its truth. They tend to despise those who do not believe. They adopt certain ritualistic (experimental) methods. They venerate certain (Nobel Prize-winning) saints. They study their scriptures. They enjoy the power and the sense of importance that science brings to them. They work and live in monastic communities called laboratories and universities, and regularly attend services, called conferences, where their high priests can be heard to preach. Finally, like most primitive religions (but unlike the mature ones), they observe the degrading practice of animal sacrifice. Like so many superstitious acolytes before them, they are convinced that such sacrifices are necessary, and they wildly exaggerate their effectiveness.

I am not opposed to science. Indeed, I am an admirer of science. I am merely pointing out, as a psychologist, that there are many psychological parallels between science and religion. Religion fulfils three huge psychological functions – it provides *meaning* (an explanation of the universe), *magic* (the alleged power through prayer and worship to influence future events) and *morality* (a guide to right and wrong behaviour). Science excels at the first two of these, providing an unprecedented understanding of the Universe and impressive technologies with which to control it. What science lacks, however, is an intrinsic morality. This is where it is weak and this is why it is dangerous. One reason why we live in such a dangerous world today is because disaffected and angry individuals and groups can gain access to the military products of science and technology.

THE DANGERS OF SCIENCE

Weapons tend to bypass our evolved internal constraints upon aggression. The longer the range of a weapon, the easier it is to avoid feeling compassion for an enemy or to fear his response. Without these emotional constraints we become more aggressive. There have always been fanatics and rebels but their power to destroy the lives of others has been limited by the power of their weapons. In the past, the influence of terrorists has been restricted by the range and power of their bows or muskets. Today, armed with nuclear or biological devices, one man can destroy a city. It is clear that science, and the products of science, need to be kept under firm moral control. Without such control science will destroy us. It brings miracles but also potential destruction for all living things.

In the recent past, the arrogance of biological science and its amoral attitude towards most forms of life, bred several generations of cold and dangerous people. They had had most of their natural human feelings of compassion and sensitivity bleached out of them by being inured to the practice of vivisection. These individuals are now dead and gone. But I can remember some of them, then very old, from my student days. Cold and sinister people – women as well as men – who believed that science was a sacred mission that gave them a right to do almost anything.

As a psychologist with a special interest in adolescence I am more than ever convinced that the way in which a teenager is taught to treat animals is a key to their adult personality generally. Such attitudes shape a whole outlook on life. Just as we now know, thanks to psychological research, that sadistic behaviour towards animals in childhood is often an indicator of sadistic and murderous behaviour towards humans later in life (many serial killers have a history of severe animal abuse), so also, I believe, the teaching of dissection and vivisection to teenagers is one of the most psychologically dangerous of all forms of instruction. Some special military forces are routinely taught to kill animals as part of their training. It helps, so the theory has it, to harden them, to make them into guiltless killers without compunction or pity. Do we want our children to be like this?

The failure by psychology to study these phenomena is itself curious. It is as if psychologists do not wish to investigate the deliberate removal of natural sensitivity and compassion, afraid of what they might discover. Yet the natural aversion to the perception of blood and injury is a very powerful and valuable feeling; so powerful that it causes medical students to faint when attending surgical operations. It is also very common. I found that, even among 15-year-old girls who had *chosen* to study biology, no less than 55 per cent said they disliked dissection and 33 per cent said such procedures actually made them feel sick or ill (these were the findings of research I did in Oxford in 1977). This indicates a strong and widespread form of behaviour that simply has been ignored by psychologists.

The general point I am making is this. Human beings are born with potentials not only to be aggressive and cruel but also with potentials for squeamishness and compassion. These last two are, surely, the foundations of morality and civilization itself. They can, however, all too easily, be stamped out by parents and teachers, and one of the quickest and surest methods of doing so, I believe, is by teaching dissection and, worse, vivisection. Dissection and vivisection in schools and colleges can take on a primitive ritual quality. Once initiated, the individual is tribally and psychologically scarred. Do we want a society full of such people?

STANLEY MILGRAM

Of course, I am not saying that all vivisectors are like psychopaths. Far from it. Normal people all too easily can be duped into doing awful things if they are told to do so by some apparent 'authority', or under the influence of peer group pressure or parental example. People like to go along with the herd. They want to be agreeable to their leaders and to their colleagues. This is fine as long as the norms of their colleagues and leaders are civilized ones, but when they become evil there is a marked tendency for most people to follow the evil trend. Look what happened in Nazi Germany. There was no shortage of ordinary people prepared to do terrible things, although, interestingly, there were also some who bravely refused and others who inwardly suffered appalling guilt, even though they followed the herd outwardly.

The Yale psychologist, Stanley Milgram, was the first to examine this phenomenon experimentally, in the 1960s. He asked volunteers to give increasingly powerful and, indeed, apparently lethal, electric shocks to students when they made mistakes in learning tasks. When the volunteers heard the students screaming and appealing for mercy many hesitated but, once they were told to continue by the white-coated scientists in charge, around 60 per cent of the volunteers continued to administer the shocks right up to the end of the scale. Of course, those apparently receiving the shocks were really unharmed actors but the volunteers did not know this. These results deeply shocked American opinion at the time. Since then, however, this research has been repeated with similar results in other countries. It seems, as Milgram concluded, that human beings can very easily behave cruelly and murderously if we find ourselves part of a wicked system.

The moral question we need to ask is: How far is animal experimentation (and speciesism generally) a wicked system? Are we already going along with evil?

CHARLES DARWIN

Charles Darwin was deeply worried about our treatment of animals. He hated cruelty. It angered him. He believed that we used animals as slaves. The moral implications of Darwinism, however, are still only beginning to sink in. Darwin proved that human beings are animals and that we are related to other animals through evolution. Thanks to Darwin, many of the huge and self-proclaimed differences between humans and animals were revealed to be no more than arrogant delusions.

Surely, if we are all related through evolution we should also be related morally. This seems to be the Darwinian message. Yet it has still not been fully accepted. Humans are given massive protection by the laws of most civilized countries while non-human animals are accorded only rudimentary rights. Non-humans can still be killed, eaten, hunted for fun, reared in appalling conditions of captivity and, of course, experimented upon. Yet science itself produces ever more evidence that non-humans can suffer pain and distress like we do. This evidence takes three main forms.

First, there is the behavioural evidence that most non-human animals will learn to avoid situations that we would find painful. Secondly, there is the anatomical evidence that shows that many species have complex nervous systems similar to ours. Thirdly, there is the biochemical evidence that indicates that in the nervous systems of all mammals, reptiles, birds and fish, are the same types of chemicals which are associated with the transmission and natural control of pain in our own. Not only 'physical' pain seems common to all higher animals but so also, as psychology indicates, are the experiences of fear, distress and boredom. We are not, as Descartes once absurdly claimed, confronted with conscious and sensitive humans on the one hand and, on the other, unconscious and unfeeling animals. There is no such gulf between us. It is a figment of human arrogance and chauvinism to assert otherwise and it is increasingly contrary to scientific evidence.

CAMPAIGNING

Let us try to be realistic about animal experimentation. All reasonable argument indicates that the same moral standards should apply to non-humans as to humans. All painients should be treated equally. So, if it is considered wrong to experiment painfully upon an unwilling human being (as happened under several twentieth-century dictatorships) then it should also be considered wrong to do so with dogs, cats, monkeys, rats or any other painient being.

But are we going to be able to stop this practice overnight? Can the massive bandwagons of the cosmetics, pharmaceutical and other scientific industries be stopped instantly? No, of course not. The huge amounts of money involved mean that this is unrealistic. However, public opinion and the growing awareness that these industries are abusing animals can have this effect gradually. Some industries themselves have shown a laudable interest in changing their own practices. Others that have not can, eventually, be shamed into doing so. After a 30-year campaign, for example, the European cosmetics industry has been forced by consumer opinion to phase out its cruel, and largely unscientific, use of animals in its laboratories. Most procedures on animals are carried out in the laboratories of big commercial firms such as these. *Such research is usually routine toxicity and other commercial testing – it is not advanced science*. Often it is carried out to meet outdated standards laid down by unaccountable Japanese or American bureaucracies or by the OECD. Bureaucracies and faceless committees that cause so much unnecessary suffering should be forced out into the open and obliged to explain themselves.

Most animal protection campaigners have, rightly, eschewed violence. It is almost always immoral to use violence in order to try to stop violence. Governments have, however, been incredibly slow to respond to public opinion and the requests of peaceful campaigners on matters of animal protection. Public opinion in Northern Europe seems to be years ahead of its governments. This is foolishness on the part of governments as it is likely to encourage unpeaceful actions. Governments often have themselves to blame for provoking lawless behaviour through their inaction and failure to listen. Nevertheless, eventually, governments will act after years of sustained political pressure involving publicity, letter writing, celebrity action, consumer boycotts, litigation, peaceful protest and scientifically well-informed lobbying. In 1997, for example, after a long campaign, a protocol was inserted into the EU Treaty itself requiring the EU and its member states to 'pay full regard to the welfare requirements of animals' in the formulation of its policies generally.

The aims of gradualist campaigners critical of animal experimentation have consistently centred upon five main questions:

1 *Are painful experiments ethical?* Increasingly, public opinion and the arguments of most professional philosophers have rejected the propriety of such research.
2 *Can humane alternative techniques be developed and used?* To an extent, this has happened. The European Centre for the Validation of Alternative Methods in Europe, although underfunded, has made some progress and so has the Interagency Coordinating Committee on the Validation of Alternative Methods in the United States. Human tissues grown in culture are now widely used.
3 *Can experiments for non-medical purposes be stopped entirely?* Inevitably, there have been disagreements about what constitutes a 'medical' purpose. Public opinion, however, makes a clear

distinction. Whereas opinions are divided on medical experimentation, around 90 per cent of adult opinion in Northern Europe has been found to be against non-medical cosmetics testing. The cruel testing of food additives and weaponry can incur approximately the same high level of condemnation.

4 *Can all pain be abolished in animal experimentation?* How to control the pain, distress and boredom of laboratory animals have become active areas of research. Techniques of analgesia, anaesthesia, euthanasia, environmental enrichment and animal care have been developed. (Of course, not all experiments need to be painful. Some psychologists once experimented to see if they could alter the direction of whales swimming in the Atlantic by sending 'telepathic' messages to them. This failed. I am fairly sure this caused the whales no pain whatsoever!)

5 *Does animal experimentation produce invalid results, anyway?* In some countries, doing research out of sheer habit or to meet unscientific or absurd bureaucratic requirements has declined, and so has the tendency to make inflated claims for the efficacy of animal procedures. There have been some tragic mistakes made by applying to humans the results of tests on other species. For example, thalidomide and penicillin reacted quite differently in laboratory species commonly in use in the past and, more recently, many non-human primates have endured substantial suffering in futile xenotransplantation research.

The so-called three Rs of reduction, refinement and replacement have been much discussed. The most important of these are replacement and refinement (where pain-reducing techniques are applied). Generally speaking, the more backward countries using laboratory animals have, at least, introduced systems of licensing of experimenters and the registration of laboratories. These are the first steps towards better control. Regular inspection is the next. Some firm restriction on pain is the basic legislative requirement and, preferably, one that can be made progressively more stringent.

THE CONCERNED EXPERIMENTER

How should the morally concerned experimenter conduct herself? If she cannot yet bring herself to stop the practice entirely, what steps can she take other than to make sure that she is as fully trained as possible in the skills of good animal care, analgesia, anaesthesia and euthanasia? A full awareness of modern alternative techniques, not involving the use of animals, is also necessary and so is a knowledge of how to plan research so that it avoids any unnecessary animal abuse. But if she is sincerely convinced that her research is really likely to provide benefits for the treatment of illness or suffering in others, she is caught in an age-old ethical dilemma: when is it ethically permissible to cause suffering to one individual in order to reduce the suffering of others?

In my opinion, the experimenter should bear in mind the following ethical rules of thumb:

1 *Speciesism is always wrong.* So try to act as though human and non-human suffering carry equal weight.
2 *The aggregation of pains or pleasures across individuals is meaningless.* So try to act as though the sufferings of the many count for no more than the sufferings of the one. Each individual matters.
3 *The primary moral concern is for the individuals who are the greatest sufferers.* Reducing the quantity of individuals who are suffering is rather pointless. It is the reduction in the *severity of pain suffered by each individual* that is the correct focus of

concern, and priority should be given to those individuals who are suffering most.

4 *It is always wrong to cause pain to 'A' merely in order to increase the pleasure of 'B'*. So inflicting even mild pain for fun or luxury, or conducting experiments for trivial purposes, or merely to advance a career, is never justified.

5 *It is always wrong, whatever the benefits, to cause severe or prolonged pain*. So, regardless of benefits, torture is never permitted. Nor are severely painful experiments.

One of the great moral disadvantages for animal experimenters is that their allegedly beneficial (for example, medical) results are always in the future. So they are trying to justify present pain by pointing to hypothetical benefits that have not yet happened. That is to say the pain is certain but the benefits are entirely uncertain. Are we, then, to allow any crank that comes along to torture animals (or humans for that matter) on *his* claim (no doubt sincerely held) that he may thereby, at some undetermined moment in the future, bring benefits to others? Is this a fair analogy?

Well, we should remember all the literally billions of animals that have died in laboratories over the last 200 years. How many of these deaths have caused any significant benefits? In retrospect, and with the benefit of hindsight, we can see that most experiments have come to nothing. Only about one in every 100 are even published in a reputable journal. Each of these papers are, on average, fully read by only about three people and only very few of these ever lead on to practical applications! So, I would estimate that no more than one hundredth of one per cent of animal experiments ever produce results that are of significant benefit to anybody, and many of these could have been done by humane alternative means anyway.

Animal experimentation, so it seems, has become a bad habit. It is perpetuated by unscientific bureaucracies and by outdated career structures. Although much of this research is carried out at the taxpayers' or shareholders' expense, most of it takes place furtively behind locked doors and is hidden from the view of a public that overwhelmingly condemns it. Can this be right?

Is some form of compromise desirable and possible between public opinion and animal researchers? In most Northern European countries animal welfarists and concerned scientists have managed to come together, to an extent, over recent decades. Many scientists in Europe genuinely want to stop the worst abuses of the past and want to find ways to meet the increasingly humane requirements of the law. Unfortunately, authorities in Japan and America, as well as the World Trade Organisation, show less interest in adopting such a civilized approach and this is very much to be regretted. In consequence, all areas of science suffer a bad image. *One of the principal reasons why children and young people do not like science is because of its cruelty to animals in laboratories.*

CONCLUSIONS

Six main points emerge from this essay:

- The most powerful argument against painful or distressing experiments on animals is the moral one. The scientific evidence is that non-human animals can suffer pain and distress in a similar manner to ourselves. Since Darwin, we have known that we are all animals. Merely because an individual comes from another species constitutes no rational grounds for causing it suffering, anymore than does a difference in gender or race. Such institutional speciesism is irrational and wrong. We should be kind to gentle aliens, too!

- Training children or students in the techniques of dissection or vivisection is psychologically dangerous and morally irresponsible.
- Most procedures on animals are carried out for routine bureaucratic purposes, often for the testing of luxury or inessential products or to cover companies against actions for damages. This is not high-minded science but runaway bureaucracy. Politicians who are not, for the most part, scientists, completely fail to grasp this.
- Many genuine scientific experiments on animals are not for strictly medical purposes. Their chances of producing any significant benefit for human or non-human beings is very small indeed. Enthusiastic experimenters, however, tend wildly to exaggerate the significance of their own research, often in order to attract funding.
- Humane alternative techniques not using live animals, such as the use of organ and tissue cultures, have been considerably developed in recent decades and governments should do far more to ensure their further development and use. For example, Magnetic Resonance Imaging techniques can be used to replace experimental brain surgery on animals.
- Democratic governments are wise to listen to public opinion on this subject because strongly felt issues carry votes. There is a strong case for far stricter controls on animal experimentation, until it can be stopped entirely. Science brings many benefits but it has also brought disasters. Secrecy surrounding animal experimentation is unjustified, especially when such research is being done allegedly in the public interest and often at the public's expense.

ACKNOWLEDGEMENT

I am grateful to Dr Maggie Jennings of the RSPCA for providing some technical information for this essay.

AFTERWORD
Tony Gilland

The contributors to this book have provided very different answers to the question of whether we should regard animal experimentation as right or wrong. Their responses reflect their different approaches to the related questions of the nature of the relationship between humans and animals; whether animals experience pain in a similar way to humans; and the meaning and content of rights. Three very different policy responses follow from the arguments that each author has made.

ABOLISH ANIMAL EXPERIMENTS

Both Tom Regan and Richard Ryder have taken a very clear stance on the question of animal experiments. They regard any such experiments as simply morally wrong. This viewpoint is based on the argument that animals experience pain in much the same way as humans do. As Regan argues, whilst animals 'lack the ability to read, write, or make moral choices', what is done to animals, 'what they experience and what they are deprived of ... matters to them.' Animals 'are the subjects of a life' and therefore have as much right to bodily integrity and life as humans do. From this perspective we cannot consistently uphold our own rights at the same time as denying the rights of animals.

The logic of this position is that society should outlaw the use of animals in research, whatever the purposes of that research. Regan is unequivocal on this: 'Vivisection is morally wrong. It should never have begun, and like all great evils, it ought to end, the sooner the better.' Regan does not say how to end vivisection, Ryder, on the other hand, makes a number of pragmatic suggestions, answering, 'No, of course not' to the question of whether the massive bandwagons of the cosmetics, pharmaceutical and other scientific industries can be stopped instantly. Consequently, Ryder is supportive of the three Rs – reduction, refinement and replacement – to the extent that they offer a step in the right direction. Additionally, he offers a number of 'ethical rules of thumb' to assist the concerned experimenter torn between alleviating human suffering and inflicting suffering on animals. Though they share a very similar analysis of the morality of animal experiments and desire the same endpoint, this is a distinction between the arguments of Ryder and the unequivocal position of Regan. The only concession that Regan is willing to consider is that 'it is conceivable [that] some uses of animals for educational purposes' (he gives the example of students observing the behaviour of rehabilitated animals returned to their natural habitat) 'might be justified.'

REGULATE FOR ANIMAL WELFARE

Mark Matfield's essay provides a fascinating account of the history of animal experimentation and the ethical issues that have concerned people over time. The willingness of the Greeks to conduct experiments on 'fully conscious animals' in the pursuit of knowledge provides a striking historical contrast to the concerned British scientists of the nineteenth century who were reluctant to perform animal experiments until the introduction of general anaesthesia in the late 1860s. Interestingly it would also seem to suggest a different, more straightforwardly human-centric attitude amongst the Greeks towards the use of animals for human advancement compared with the dominant view held by scientists today. The latter view, through advocating the importance of the three Rs and strict regulation of experiments, implies a trade-off between the interests of animals and humans and attaches a greater weight to the interests of animals than the Greeks might have done. As Matfield argues: 'The degree of cruelty and suffering involved would have been intolerable by modern standards. These experiments were performed on fully conscious animals.' Whether modern-day regulations simply reflect greater technical capabilities for catering for the welfare of animals – and in turn improving test results – or a shift to a less human-centric viewpoint is an interesting question.

Matfield appears comfortable with, and supportive of, the regulatory position that we have arrived at today. He does not equate human suffering with that of animals and clearly believes that the two are sufficiently distinct for the benefits that animal experiments bring to humans to justify the suffering caused to animals. However, though Matfield writes proudly about the achievements of medical science, he is also keen to underline the importance he attaches to animal welfare and the approach dictated by the three Rs. His position would appear to be that animal experiments are justifiable (and have contributed to significant human achievements) but must be conducted humanely with a view to minimizing any pain or suffering that might be experienced by the animals involved.

Others within the scientific community, such as the Royal Society, the UK's premier scientific body, adopt a very similar position to Matfield. It is probably the mainstream view amongst scientists, though some are highly critical of existing regulations being too stringent and unnecessarily obstructive. It also seems likely that the adoption of the three Rs and the extent of today's regulatory requirements reflect fears about public disapproval of animal experimentation, as well as concern for the humane treatment of animals. Matfield notes how 'fear of being targeted by animal rights extremists' has made animal researchers 'unwilling to engage in the public debate' and 'reluctant to be more open'. He does believe, however, that greater openness on the part of scientists would lead to greater public acceptance. From this viewpoint, there is no particular need to change the existing regulatory and policy framework governing animal experimentation.

PRIORITIZE GOOD SCIENCE

Stuart Derbyshire, on the other hand, not only vehemently objects to the abolitionist position of Regan and Ryder, but is also highly critical of what he believes to be misplaced concessions made by scientists and politicians keen to assuage public concerns and reflect the influence of the animal rights movement. Derbyshire is not only concerned with the danger that unnecessary regulations might obstruct useful scientific endeavours, he is also concerned about the way in which the adoption of the three Rs projects 'a lowlife opinion of animal researchers' and encourages 'the notion that animal experiments are problematic'. He believes that well cared-for animals generally make the best subjects for scientific research, but that what should govern laboratory practices is what produces the best results not the welfare of animals as an end in itself. From his perspective, to argue anything else is essentially dishonest: 'Giving AIDS and other diseases to animals, carrying out experimental surgeries and infusing untested drugs hardly sound like procedures aimed at protecting the animals' welfare. Professed concern for the welfare of laboratory animals is simply inconsistent with the reality of laboratory experiments that almost invariably result in distress and death for the animal.'

Derbyshire's arguments are based upon what he describes as an 'uncompromisingly human-centric' outlook and his understanding of the distinction between humans and animals. According to Derbyshire, only humans are capable of making conscious choices, and only humans are truly subjects. Therefore only humans are capable of possessing rights. Derbyshire would undoubtedly support the removal of many of the restrictions of animal experiments and prefer to see society believe in the endeavours of science and the responsibility of scientists. His concern would appear to be not only with animal experiments but also about how contemporary views about this issue reflect a negative view of human beings.

The essays in this book suggest that the issues raised by the question 'Animal Experimentation: Right or Wrong?' present practical questions for policy-makers, but also bring to our attention broader moral questions about the meaning of rights and responsibilities, and the nature of the human experience.

DEBATING MATTERS

Institute of Ideas
Expanding the Boundaries of Public Debate

If you have found this book interesting, and agree that 'debating matters', you can find out more about the Institute of Ideas and our programme of live conferences and debates by visiting our website **www.instituteofideas.com**.
Alternatively you can email **info@instituteofideas.com**
or call 020 7269 9220 to receive a full programme of events and information about joining the Institute of Ideas.

Other titles available in this series:

DEBATING MATTERS

Institute of Ideas
Expanding the Boundaries of Public Debate

ABORTION:

WHOSE RIGHT?

Currently around 180 000 British women terminate pregnancies each year – far more than the politicians who passed the Abortion Act in 1967 intended. Should the law be made more liberal to reflect demand for abortion? Is the problem that in Britain, women still do not have the 'right to choose'? Or is it too easy for women to 'take the life' of their 'unborn children'? What role should doctors play in the abortion decision?

Contrasting answers are presented in this book by:

- Ann Furedi, director of communications, British Pregnancy Advisory Service
- Mary Kenny, journalist and writer
- Theodore Darymple, GP and author of *Mass Listeria: The Meaning of Health Scares* and *An Intelligent Person's Guide to Medicine*.
- Emily Jackson, Lecturer in Law, London School of Economics
- Helen Watt, director, Linacre Centre for Healthcare Ethics.

REALITY TV:

HOW REAL IS REAL?

What is reality TV, and how real is it anyway? From gameshows such as *Big Brother* to docusoaps and even history programmes, television seems to be turning its attentions onto 'real people'. Does this mean that television is becoming more democratic, or is reality TV a fad that has had its day? Does reality TV reflect society as it really is, or merely manufacture disposable celebrities?

Contrasting views come from:

- Christopher Dunkley, television critic for the *Financial Times*
- Dr Graham Barnfield, Lecturer in Journalism,
  The Surrey Institute of Art & Design
- Victoria Mapplebeck, TV producer of the TV shows
  Smart Hearts and *Meet the Kilshaws*
- Bernard Clark, documentary maker.

ETHICAL TOURISM:

The idea of 'responsible tourism' has grown in popularity over the past decade. But who benefits from this notion? Should the behaviour of travellers come under scrutiny? What are the consequences of this new etiquette for the travelling experience? Can we make a positive difference if we change the way the travel?

Contrasting responses to these questions come from:

- Dea Birkett, columnist for *The Guardian* and author of *Amazonian*
- Jim Butcher, Senior Lecturer, Department of Geography and Tourism, Canterbury Christ Church University College
- Paul Goldstein, Marketing Manager, Exodus Travel
- Dr Harold Goodwin, Director of the Centre for Responsible Tourism at the University of Greenwich
- Kirk Leech, Assistant Director of the youth charity Worldwrite

SCIENCE:

CAN WE TRUST THE EXPERTS?

Controversies surrounding a plethora of issues, from the
MMR vaccine to mobile phones, from BSE to genetically-
modified foods, have led many to ask how the public's faith
in government advice can be restored. At the heart of the
matter is the role of the expert and the question of whose
opinion to trust.

In this book, prominent participants in the debate tell us
their views:

- Bill Durodié, who researches risk and precaution at New College,
  Oxford University
- Dr Ian Gibson MP, Chairman of the Parliamentary Office of
  Science and Technology
- Dr Sue Mayer, Executive Director of Genewatch UK
- Dr Doug Parr, Chief Scientist for Greenpeace UK.

NATURE'S REVENGE?

HURRICANES, FLOODS AND CLIMATE CHANGE

Politicians and the media rarely miss the opportunity that hurricanes or extensive flooding provide to warn us of the potential dangers of global warming. This is nature's 'wake-up call' we are told and we must adjust our lifestyles.

This book brings together scientific experts and social commentators to debate whether we really are seeing 'nature's revenge':

- Dr Mike Hulme, Executive Director of the Tyndall Centre for Climate Change Research
- Julian Morris, Director of International Policy Network
- Professor Peter Sammonds, who researches natural hazards at University College London
- Charles Secrett, Executive Director of Friends of the Earth.

DESIGNER BABIES:

WHERE SHOULD WE DRAW THE LINE?

Science fiction has been preoccupied with technologies to control the characteristics of our children since the publication of Aldous Huxley's *Brave New World*. Current arguments about 'designer babies' almost always demand that lines should be drawn and regulations tightened. But where should regulation stop and patient choice in the use of reproductive technology begin?

The following contributors set out their arguments:

- Juliet Tizzard, advocate for advances in reproductive medicine
- Professor John Harris, ethicist
- Veronica English and Ann Sommerville of the British Medical Association
- Josephine Quintavalle, pro-life spokesperson
- Agnes Fletcher, disability rights campaigner.

ALTERNATIVE MEDICINE:

SHOULD WE SWALLOW IT?

Complementary and Alternative Medicine (CAM) is an increasingly acceptable part of the repertory of healthcare professionals and is becoming more and more popular with the public. It seems that CAM has come of age – but should we swallow it?

Contributors to this book make the case for and against CAM:

- Michael Fitzpatrick, General Practitioner and author of *The Tyranny of Health*
- Brid Hehir, nurse and regular contributor to the nursing press
- Sarah Cant, Senior Lecturer in Applied Social Sciences
- Anthony Campbell, Emeritus Consultant Physician at The Royal London Homeopathic Hospital
- Michael Fox, Chief Executive of the Foundation for Integrated Medicine.

TEENAGE SEX:

WHAT SHOULD SCHOOLS TEACH CHILDREN?

Under New Labour, sex education is a big priority. New policies in this area are guaranteed to generate a furious debate. 'Pro-family' groups contend that young people are not given a clear message about right and wrong. Others argue there is still too little sex education. And some worry that all too often sex education stigmatizes sex. So what should schools teach children about sex?

Contrasting approaches to this topical and contentious question are debated by:

- Simon Blake, Director of the Sex Education Forum
- Peter Hitchens, a columnist for the *Mail on Sunday*
- Janine Jolly, health promotion specialist
- David J. Landry, of the US based Alan Guttmacher Institute
- Peter Tatchell, human rights activist
- Stuart Waiton, journalist and researcher.

COMPENSATION CRAZY:

DO WE BLAME AND CLAIM TOO MUCH?

Big compensation pay-outs make the headlines. New style 'claims centres' advertise for accident victims promising 'where there's blame, there's a claim'. Many commentators fear Britain is experiencing a US-style compensation craze. But what's wrong with holding employers and businesses to account? Or are we now too ready to reach for our lawyers and to find someone to blame when things go wrong?

These questions and more are discussed by:

- Ian Walker, personal injury litigator
- Tracey Brown, risk analyst
- John Peysner, Professor of civil litigation
- Daniel Lloyd, lawyer.

THE INTERNET:

BRAVE NEW WORLD?

Over the last decade, the internet has become part of everyday life. Along with the benefits however, come fears of unbridled hate speech and pornography. More profoundly, perhaps, there is a worry that virtual relationships will replace the real thing, creating a sterile, soulless society. How much is the internet changing the world?

Contrasting answers come from:

- Peter Watts, lecturer in Applied Social Sciences at Canterbury Christ Church University College
- Chris Evans, lecturer in Multimedia Computing and the founder of Internet Freedom
- Ruth Dixon, Deputy Chief Executive of the Internet Watch Foundation
- Helene Guldberg and Sandy Starr, Managing Editor and Press Officer respectively at the online publication *spiked*.

ART:

WHAT IS IT GOOD FOR?

Art seems to be more popular and fashionable today than ever before. At the same time, art is changing, and much contemporary work does not fit into the categories of the past. Is 'conceptual' work art at all? Should artists learn a traditional craft before their work is considered valuable? Can we learn to love art, or must we take it or leave it?

These questions and more are discussed by:

- David Lee, art critic and editor of *The Jackdaw*
- Ricardo P. Floodsky, editor of artrumour.com
- Andrew McIlroy, an international advisor on cultural policy
- Sacha Craddock, an art teacher and critic
- Pavel Buchler, Professor of Art and Design at Manchester Metropolitan University
- Aidan Campbell, art critic and author.